Johannes Wippler

Micromechanical Finite Element Simulations of Crack Propagation in Silicon Nitride

Schriftenreihe
Kontinuumsmechanik im Maschinenbau
Band 3

Karlsruher Institut für Technologie
Institut für Technische Mechanik
Bereich Kontinuumsmechanik
Hrsg. Prof. Dr.-Ing.habil. Thomas Böhlke

Eine Übersicht über alle bisher in dieser Schriftenreihe erschienenen Bände finden Sie am Ende des Buchs.

Micromechanical Finite Element Simulations of Crack Propagation in Silicon Nitride

by
Johannes Wippler

Dissertation, Karlsruher Institut für Technologie
Fakultät für Maschinenbau
Tag der mündlichen Prüfung: 14. Dezember 2011

Impressum

Karlsruher Institut für Technologie (KIT)
KIT Scientific Publishing
Straße am Forum 2
D-76131 Karlsruhe
www.ksp.kit.edu

KIT – Universität des Landes Baden-Württemberg und nationales
Forschungszentrum in der Helmholtz-Gemeinschaft

KIT Scientific Publishing 2012
Print on Demand

ISSN: 2192-693X
ISBN: 978-3-86644-818-6

Micromechanical Finite Element Simulations of Crack Propagation in Silicon Nitride

Zur Erlangung des akademischen Grades
Doktor der Ingenieurwissenschaften
der Fakultät für Maschinenbau
Karlsruher Institut für Technologie

genehmigte
Dissertation
von

Johannes Wippler

Tag der mündlichen Prüfung: 14.12.2011
Hauptreferent: Prof. Dr.-Ing. Thomas Böhlke
Korreferent: Prof. Dr. Michael J. Hoffmann

Logic merely sanctions the conquests of intuition.

Jaques Hadamard

Zusammenfassung

Siliciumnitrid wird in der Industrie zum Beispiel für Wendeschneidplatten oder Formwalzen verwendet. Sehr wichtig für diese herausfordernden Anwendungen ist die besondere Kombination von hoher Festigkeit und Bruchzähigkeit. Die Mikrostruktur aus nadelartigen β-Si_3N_4-Körnern in einer Matrix aus oxidnitridischem Glas sorgt durch die gestreckten und – im Vergleich zu den Korngrenzen – festen Körnern für Rissumlenkung und damit für Zähigkeit.

Für ein besseres Verständnis dieses komplexen Materialverhaltens auf der Mikroskala wurde ein Gefügemodell sowie Materialmodelle für die thermoelastischen Eigenschaften und das Bruchverhalten implementiert, welche die Erzeugung von dreidimensionalen Einheitszellen für Finite-Element-Simulation erlauben. Diese wurden für mikromechanische Simulationen eingesetzt. Dadurch konnte einer der wichtigsten Verstärkungsmechanismen in Siliciumnitrid – die Ausbildung von elastischen Brückenkörnern – nachgewiesen sowie ihr Versagen beobachtet werden. Die Berechnung von effektiven Belastungskurven ermöglichte es, Aussagen über den Einfluss der Spannungsmehrachsigkeit und der Gefügemorphologie auf das Bruchverhalten von Siliciumnitrid zu treffen.

Diese mikromechansischen Berechnungen sind aber für eine praktische Anwendung in Bauteilberechnungn wesentlich zu aufwändig. Also wurde ein effektives Bruchmodell entwickelt, das einen Vergleich der Ergebnisse auf der Mikroebene mit experimentellen Befunden auf der Makroebene ermöglicht. So konnten *R*-Kurven-Experimente durch Simulationen auf Basis der mikromechanischen Ergebnisse nachvollzogen werden.

Diese Arbeit ermöglicht ein besseres Verständnis des Bruchverhaltens von Siliciumnitrid und ebnet damit den Weg für neue Anwendungen.

Summary

Silicon nitride is used for cutting inserts or forming rolls. Crucial for these challenging applications is the specific combination of high strength and fracture toughness. The microstructure consists of column-like β-Si_3N_4 grains in a matrix of oxynitride glass. The elongated and – compared with the grain interfaces – strong grains support the crack path deviation and, therefore, the fracture toughness.

A better understanding of this complex material behavior on the microscale can be achieved by the implementation of three-dimensional unit cells for finite element simulations, which include a structural model, the thermoelastic properties as well as the fracture behavior. They have been used for micromechanical calculations. One key result is that one of the most important reinforcement mechanisms – the evolution of elastic bridging grains – could be substantiated and observed during their failure. The determination of effective load paths allowed for the characterization with respect to the influences of the stress triaxiality and structural morphology on the fracture behavior of silicon nitride.

These micromechanical simulations are far too costly for a practical application in structural design. Hence, an effective fracture model has been implemented, which allows for a comparison of the findings on the microscale with experimental results. So, based on the micromechanical results, R-curve experiments have been reproduced.

This thesis improves the understanding for the fracture behavior of silicon nitride and clears the way for new applications.

Acknowledgment

Special thanks go to my doctoral thesis supervisor Prof. Thomas Böhlke. From him I learned the basics of continuum mechanics. Additionally, I thank him for academic freedom and for scientific rigor, which I really appreciate and which I tried make the best use of for my work.

I thank Prof. Michael J. Hoffmann for his engagement as co-referee for the thesis and for valuable inspirations on the wide field of silicon nitride research.

I very much appreciated the scientific support from Dr. Theo Fett, Dr. Stefan Fünfschilling, Dr. Marco Riva and Dr. Rainer Oberacker from the Institute for Applied Materials – Institute for Ceramics in Mechanical Engineering, Prof. Alexander Wanner from the Institute for Applied Materials – Materials Science and Engineering at the Karlsruhe Institute of Technology, Prof. Frank Mücklich from the University of Saarland, Dr. Matthias Herrmann from the Fraunhofer Institute in Dresden, Dr. Tanja Lube from the Montan University Leoben, Prof. Stuart Hampshire from the University of Limerick, Dr. Paul F. Becher from Oakridge National Lab and Prof. Yujie Wei from the Massachusetts Institute of Technology. Thank you for all your support!

A very special thank you goes to my highly appreciated colleague and friend Dr.-Ing., Dipl.-math. techn. Felix Fritzen, who was always prepared to answer my questions and to give me further inspiration on mechanics and programming. You did so much for this work!

Thank you as well for the impressive technical support from our software providers Simpleware, especially by Ross Cotton and from Abaqus, especially by Dr. Fengming Zhou.

Thank you so much for the support and your patience, Dr. Paul Weber and Dr. Horst Gernert from the Steinbach Computation Center.

I will never forget the friendly and completely informal support for experimental issues from Barthel Brylka, Rolf Stober, Claudius Wörner, Dr. Kay Weidenmann and Robert Welker.

Heartfelt thanks to our secretaries Helga Betsarkis and Ute Schlumberger-Maas for the proof reading and all the administrative support.

Extremely helpful for me as a scientist and as well as a person was the fruitful collaboration with my eager students, who have written their theses under my supervision or which helped me as assistants. They highly contributed to my work due to their critical questions, their candity and most of all by their good work. Thank you, Thomas Altmayer, Jan Breitschuh, Michael Bürckert, Aswin Chandar, Nicole Fischer, Andreas Flock, Vladimir Josanovic, Johannes Laidig, Qiuning Luo, Benjamin Nothdurft, Mrudula Nujella, Stephan Prieß, Jing Qian and Rebecca Schaaf.

Further thanks go to my family, to all my colleagues and friends and, especially, to my love Ilona for the final corrections and her clever suggestions.

And finally, this acknowledgment would not be complete without an appraisal of the acoustic inspiration by Bach, Beethoven and www.bassdrive.com for the absolutely kneep-deep vibes.

Contents

Chapter 1

Introduction

1.1 Motivation

In the last decades, the development of high performance materials for functional as well as for structural applications has played a major role in science. Examples of such materials for structural applications are high strength steels or super alloys, which have to be cut or formed, see Figure 1.1 a). It is clear that an efficient treatment of these stiff and strong materials requires tools (Figure 1.1 b), which are stiffer and stronger than the workpieces at high temperatures. Additionally, they require a long lifetime, since the production has to be as cheap as possible due to the increasing cost pressure. So, the toughness and the wear resistance as well as the thermal shock resistance play the key roles in the selection of tool materials.

The sound combination of these material properties is a challenging task for material science. Ceramics turned out to provide this property profile in many cases. Due to their atomic structure with mainly covalent binding they are strong and stiff, but also brittle. This can lead to abrupt failure and thus to production interruption

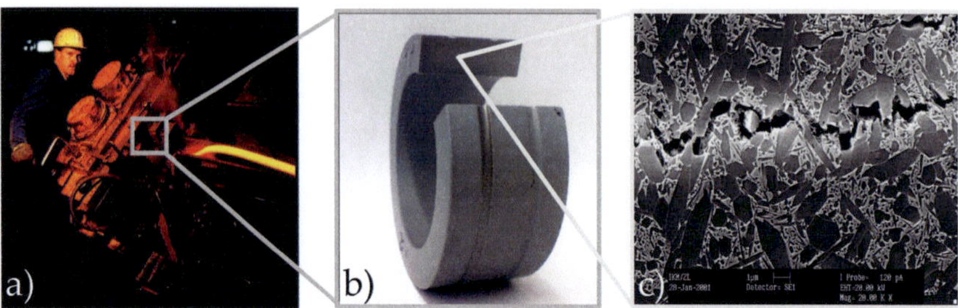

Figure 1.1: Silicon nitride and its application in demanding industrial processes: a) Production of high-strength steel wires, b) ceramic roller tool for a process shown in a), and c) silicon nitride microstructure with a crack that is deviating around the grains (Satet, 2002).

or even worse consequences. So, the toughness has to be described as exact as possible to allow secure application under complex conditions. From the view of a material scientist, especially the microstructure together with the intergranular properties play a very important role for the overall strength and toughness of engineering ceramics. Figure 1.1 c) shows the microstructure of silicon nitride, which provides the features that lead to tough and strong materials. It can be seen that the crack in this material cannot propagate on a straight line due to the strong grains, which are debonding or cracking and thus dissipating energy. This leads to high critical energy release rates and, therefore, to a high fracture toughness.

This microstructure is the main reason for the prominent feature combination of silicon nitride, which makes it the first choice when an optimal combination of high strength, wear resistance, and, most important fracture toughness, is needed. The relatively low elastic stiffness and its low thermal expansion coefficient together with the high strength and thermal conductivity lead to an excellent thermal shock resistance. Altogether, this is a unique property constellation, which makes silicon nitride irreplaceable as a tool material for industrial high performance applications such as the turning of cast iron with discontinuous cut or the hot wire rolling of high strength steel wires, see, Figure 1.1 a) and b). Here, the tools have to resist extreme thermal and mechanical challenges due to the dynamic loading at extreme magnitudes.

Considering these facts it appears highly desirable to be capable of predicting the properties of silicon nitride. In this sense, a huge amount of scientific work has been done. It was found that the complex microstructure with large grains of high aspect ratio together with an appropriate grain boundary toughness plays a crucial role in order to achieve an ideal combination of strength and toughness.

Compared to the great majority of experimental material assessment, which is mainly paired with linear elastic fracture mechanics approaches, the penetration of the topic by finite element simulations is relatively small. This is due to the challenges that are linked to a sound modeling of the material, which has to incorporate the geometry as well as the material behavior on the microscale into one model. Every single contribution is linked to considerable theoretical issues, which have not yet been undertaken in this voluminous combination to the best knowledge of the author.

1.2 State of the Art

The high level of knowledge concerning the production and the experimental assessment of silicon nitride becomes obvious when looking at the huge variety of publications. The most relevant considerations for this work will be addressed briefly in this section. A more detailed inspection will follow at the beginning of the chapters.

One of the most important characteristics of silicon nitride have been observed by Lange (1973). His observation of the fracture surface of double cantilever samples showed that the high fracture toughness can be related to the large grains with high aspect ratios. From that time onwards, a considerable amount of work has been undertaken to make the high potential of the structural reinforcement useful. To name only a few, the work of Ohji et al. (1995) and Becher et al. (1998) have to be mentioned. They found that special distributions of grain shape and size are preferable for toughening.

Especially important for this work is the notion of elastic bridging grains, which reinforce the material in the early stages of the fracture process. Those large grains are linking the flanks of cracks without debonding and, therefore, they toughen the material significantly. This effect is the main source for the high fracture toughness and the extremely steep rising R-curve, which is observed in fracture experiments on silicon nitride as outlined in, e.g., Fünfschilling et al. (2011). Here, as well, thermal residual stresses come into play, which arise during the cool-down after the sinter process. They support the bonding of the grains and, thus, the toughness of silicon nitride, as found by Peterson and Tien (1995).

In order to understand and to exploit this effect, considerable efforts have been undertaken. Here, both the sintering conditions as well as the chemistry of the intergranular phase have been tuned in order to obtain an optimum either for toughness, strength, or a good compromise of both (Sun et al., 1998; Peillon and Thevenot, 2002; Kruzic et al., 2008).

It is highly desirable to determine how the effective properties on the length scale of structural applications arise from the constellation on the microlevel. Considering the complexity of morphology, elasticity, and fracture behavior, it is reasonable to use a unit cell with the morphology, material and interface properties.

Fundamental works on this field are Suquet (1985) and Maugin (1992), where the basic concepts are explained and applied and which have been developed further with respect to fracture mechanics by Verhoosel et al. (2010).

1.3 Concept and Overview

The main body of this work is divided into three parts, in which the different features of the proposed model for silicon nitride are addressed: Microstructure, thermoelastic properties, and finally – the main topic of the work – fracture of silicon nitride on the microscale and determination of the effective properties. The structure is outlined in Figure 1.2: On one side, a microstructure generation is necessary due to the currently invincible challenges, which are linked to experimental access to the three-dimensional microstructures. The synthetic geometry data is used for the generation of finite-element meshes. On the other hand, the numerous material features for the thermoelasticity and the trans- and intergranular fracture have to be incorporated. All results have been combined to complex finite element models. Those simulations are subsequently used for the determination of the effective properties.

Consequently, the second chapter is on the algorithmic creation of the microstructure, which tries to represent the most important geometrical properties of silicon nitride. This structural model will be used throughout the work in order to provide geometries for the finite element models.

The third chapter introduces all thermoelastic properties on the microscale and incorporates them into a conclusive framework for finite element simulations. The model consists of detailed and experimentally motivated constitutive assumptions on the microscale concerning the elastic stiffness and the thermal expansion behavior. This information is used to determine the effective thermoelastic behavior and allows to account for the thermal residual stresses.

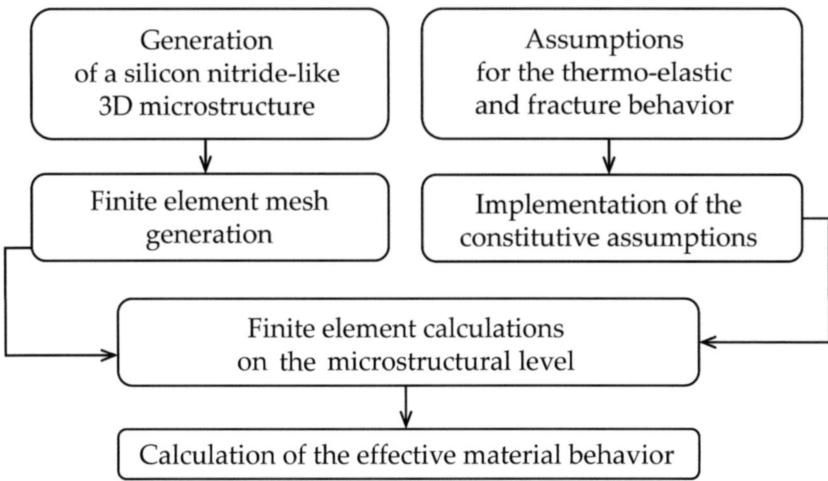

Figure 1.2: Conceptual diagram for the chosen homogenization approach. The geometry generation and the material and interface properties are merged into complex microstructural finite-element simulations, which allow for the determination of the effective properties.

The fourth chapter incorporates the results from the precedent chapters into one micromechanical finite element framework combining all information: the synthetic microstructures, thermoelastic properties, and detailed assumptions about the fracture behavior on the microscale, in particular, the fracture of the phases and the interface behavior between the grains. With these assumptions a procedure for the determination of the effective properties is proposed. For the effective fracture behavior a simplified model is suggested that can be applied to macroscopic simulations. This is examplified on an important experiment: The four-point bend test on edge notched beams. In order to prove the concept, the evaluation of the macroscopic fracture simulations will be carried out with approved fracture mechanics concepts.

The final chapter discusses the main findings of the work and provides several options for a further development and applications of the model.

1.4 Notation and Nomenclature

A direct tensor notation is preferred throughout the text. Vectors and second-order tensors are denoted by lowercase and uppercase bold letters, e.g., a and A, respectively. A linear mapping of second-order tensors by a fourth-order tensor $\mathbb{C}$ is written as $A = \mathbb{C}[B]$. The scalar product and the dyadic product between vectors and tensors are denoted by $a \cdot b$ and $a \otimes b$, respectively. The brackets $\langle \cdot \rangle$, e.g., $\langle \sigma \rangle$, indicate ensemble averaging, which can be identified with volume averages in the infinite volume limit for ergodic media. The effective quantities are denoted by a bar, e.g., $\bar{\mathbb{C}}$. The tensor I is the identity on vectors. The identity on symmetric second-order tensors is represented $\mathbb{I}^S$. All tensorial quantities are embedded in the three-dimensional Euclidean space $\mathbb{R}^3$.

The following list of variables compiles all relevant quantities used in this thesis.

It is organized in scalar and tensorial variables and sets. At the end a list of frequently used sub- and superscripts is included.

If a symbol has multiple meanings it is used in mutually exclusive contexts, so that confusion should be avoided. The mentioned quantities are always presented in their basic form and can be concretized by scripts or other symbolic extensions.

Scalar Variables

a	Crack length
A	(Crack) area
$\mathcal{A}$	Aspect ratio
B	Grain thickness
c	Volume fraction, scatter constant, speed of sound
C	Elastic continuum stiffness constant
d	Euclidean distance
$\mathcal{D}$	Dissipation
E	Young's modulus
f	Coefficient of friction, auxialliary constant
F	Force
G	Shear modulus
$\mathcal{G}$	Energy release rate
H	Degradation constant
k	Constant
K	Bulk modulus, interface stiffness component, stress intensity factor
L	Grain length
ℓ	Norm symbol, crack length
$\mathcal{L}$	Lagrangian function
p	Norm exponent, hydrostatic pressure
$\wp$	Accumulated degradational separation or strain
$\mathcal{P}$	Stress power
q	Degradation traction or stress
r	Coefficient, random number
s	Coefficient, initial interface strength
S	Compliance
$\mathcal{S}$	Degradation energy
t	Coefficient, pseudo-time, traction vector component
T	Temperature
$\mathcal{T}$	Triaxiality
u	Displacement vector component

V	Volume
v	Velocity
w	Edge length of the unit cell
y	Lattice distance
Y	Geometry function
α	Thermal expansion coefficient, normalized crack length
$\dot{\gamma}$	Lagrangian multiplier
ε	Infinitesimal strain
ϑ	Polar angle
λ	Lamé constant
μ	Lode parameter
ν	Poisson's ratio
ρ	Density
σ	Stress tensor component or stress-like variable
τ	Norm of tangential traction
φ	Azimuthal angle
ϕ	Fracture criterion
ψ	Helmholtz free energy
ω	Rotation angle

Sets

$\mathcal{D}$	Grain dimensions
$\mathcal{E}$	Prism edges
$\mathcal{F}$	Prism faces
$\mathcal{G}$	Grain (β-Si_3N_4)
$\mathcal{I}$	Interface (grain boundary)
$\mathcal{L}$	Contact indicator
$\mathcal{M}$	Matrix (glassy phase)
$\mathcal{P}$	Projections
$\mathcal{V}$	Prism vertices

Vectors and Tensors

A	Coefficient matrix
B	Second order basis tenor
$\mathbb{C}$	Elastic continuum stiffness tensor
d	Unit normal vector on unit sphere, separation vector
e	Cartesian unit vector
H	Displacement gradient
J	Jacobian
k	Slope tensor
K	Elastic interface stiffness tensor
m	Prism edge vectors
n	Normal vector
p	Point, principal vector
P	Principal projector
$\mathbb{P}$	Fourth-order projector
Q	Orientation tensor
s	Shift vector for grain origin translation
S	Elastic interface compliance tensor
$\mathbb{S}$	Elastic continuum compliance tensor
t	Traction vector
u	Displacement vector
v	Vertice vector, triangle edge vector, polynomial vector
w	Displacement fluctuation vector
x	Location
α	Thermal expansion tensor
$\dot{\gamma}$	Lagrangian multiplier vector
ε	Infinitesimal strain tensor
δ	Separation vector in tangential space
Ξ	Coupling tensor
σ	Cauchy stress tensor
τ	Traction vector in tangential space
ϕ	Vectorized fracture criterion

Sub- and Superscripts

0	Initial
acc	Accumulated
c	Crack or critical
d	Damage
eq	Equivalent
ga	Growing around
int	Intersection
m	Mean intersection point
n	Normal direction
o	Overlap
p	Periodic copy
pr	Principal
pen	Penalty
R	Resistance
t	Tangential direction
th	Thermal
μ	Micro

Chapter 2

Microstructure Generation[☆]

2.1 Introduction

Silicon nitride is a structural reinforced ceramic, which is used for high performance applications due to its good compromise of stiffness, strength, and toughness. This interesting property profile is related to different aspects of the material on the microlevel. Its strength and fracture toughness strongly depend on two features: The grain size-grain shape distribution and the properties of the grain boundary films are equally relevant for the effective mechanical material properties.

The observation of the structure-property relationship is a long-term field of research. Lange (1973) was one of the first to the knowledge of the author who investigated a relation between fracture energy and the microstructure of hot-pressed silicon nitride. He reported results from double cantilever specimens, which clearly indicate a correlation between the measured fracture energy

---

and the detected elongated grains in β-rich silicon nitride compositions. It has been observed that the samples, which dissipated the most fracture energy had a significantly increased specific surface with deep holes from pulled-out grains of high aspect ratio.

Ohji et al. (1995) and Becher et al. (1998) found that especially a bimodal distribution of grain sizes with an anisotropic distribution of the grain axes is increasing the fracture toughness, whereas large and elongated grains can act as bridges between the crack surfaces, so that cracks have to propagate under circumvention of the grains.

Different ways of achieving such microstructures have been examined. An example is given in Peillon and Thevenot (2002), where it was found that longer sintering times lead to improvement of the fracture toughness compared to seeding with big β-grains, because the seeding can have unfavorable effects on the densification of the material. With increased seeding times and, thereby, longer periods of natural grain growth, fracture toughness could be improved by up to 30% with respect to the reference material.

In Sun et al. (1998) the influence of yttria and alumina additives have been examined with the result that a high Y:Al ratio is increasing the fracture resistance due to large an elongated grains with a relatively low debonding stress, which supports crack path deviation.

Different approaches to the simulation of silicon nitride microstructure evolution have been documented. An important contribution is the implementation of the anisotropic Ostwald ripening for faceted crystals by Kitayama et al. (1998a), the α- to β-transition (Kitayama et al., 1998b) and the shape evolution of a single grain in Kitayama et al. (2000). Here, not the microstructure itself was created, such that the most important results in the light of this work are the distributions of grain lengths, widths, and aspect ratios. They are in good agreement with experimental observations and with the presented results of the synthetic structure generation.

The effective mechanical properties of a polycrystalline material can be calculated from information on the microlevel. A classical approach is to use a unit cell that is assumed to represent the geometrical and material information of the considered bulk material. Given the statistical "representativity", the unit cell is often called a representative volume element (RVE). The best way for creating such a unit cell is the direct usage of experimental data. Borbély et al. (2006) have used microtomographic observations on particle-reinforced metal-matrix composites (aluminum as matrix with 20% alumina particles as reinforcement) for the construction of finite element meshes.

In case of silicon nitride, this approach is not feasible due to the extremely challenging preparation techniques. Therefore, it is technically not possible to use a sequence of scanning electron microscope (SEM) images for the in-depth information on the silicon nitride geometry. The required slice thickness lies in the sub-micrometer domain due to the given grain sizes and cannot be delivered by recent preparation techniques. The application of image-giving techniques like magnetic resonance tomography (MRT) or micro computer tomography (μ-CT) is impossible due to the low phase contrast and the very small intrinsic dimensions, which are beneath the accessible resolution. First investigations of image stack acquisition by a combination of electron backscatter diffraction (EBSD) for the image acquisition and focused ion beam sections (FIB) for the in-depth segmentation have been undertaken. However, they are yet too expensive for an efficient geometry reconstruction.

Thus, a different way had to be chosen, which was inspired by the sequential adsorbtion technique. It allows for an algorithmic creation of a structure, which is similar to experimental observations. Louis and Gokhale (1996) created a synthetic microstructure of a polymer matrix composite with spherical inclusions of constant size. The geometry was used to obtain a self-consistent model for the electrical conductivity. Tschopp et al. (2008) used two-dimensional distributions of ellipses with different aspect ratios, which have been assembled into representative area elements for an examination of image analysis methods.

The presented algorithm for silicon nitride-like structure combines the sequential adsorbtion technique, see, e.g., Cooper (1988), with particle growth and subsequent steric hindrance and is based on experimental observations on the grain growth processes. It has to be emphasized that the algorithm does not consider any chemical or thermodynamical relations.

2.2 Observation and Implementation

2.2.1 Overview

The main idea of the structural model is that the silicon nitride-like structure is arising from statistical seeded locations and orientations of hexagonal prisms, which will be called "grains" in the context of this work. The grains are seeded in an adjustable number of steps and are thought to be growing isotropically until steric hindrance. The cases of pinning require specific considerations, which will be given later on.

After calculating the "exact" structure, the determination of the voxel structure, which is used for the finite element discretization, allows for certain smaller refinements like the angularity of the grains and the matrix volume fraction of the structure.

2.2.2 Geometric Quantities

On one hand, grains, i.e., hexagonal prisms, can be regarded as sets of vertices, edges and planes. An equivalent representation is provided by the norm concept, such that grain-like contours can be described by the norm $\ell^{\mathrm{max}} \leqslant 1$, which will be introduced in Eq. (2.8).

Both concepts use the grain origin $\boldsymbol{x}_0$, the unit normal vectors of the hexagonal prism and its dimensions B and L for the breadth and length, respectively (Figure 2.1). The orthogonal orientation tensor $\boldsymbol{Q}$ is given by the angle-axis representation

$$\boldsymbol{Q} = \boldsymbol{Q}(\boldsymbol{d},\omega) = \boldsymbol{d} \otimes \boldsymbol{d} + (\boldsymbol{I} - \boldsymbol{d} \otimes \boldsymbol{d})\cos(\omega) - \boldsymbol{\epsilon}[\boldsymbol{d}]\sin(\omega) \qquad (2.1)$$

with the rotation angle ω. The rotation axis $\boldsymbol{d}$ represents the normal vector on the unit sphere with

$$\boldsymbol{d} = \boldsymbol{d}(\vartheta,\varphi) = \sin(\vartheta)\cos(\varphi)\,\boldsymbol{e}_1 + \sin(\vartheta)\sin(\varphi)\,\boldsymbol{e}_2 + \cos(\vartheta)\boldsymbol{e}_3, \qquad (2.2)$$

where $\boldsymbol{I}$ is the identity tensor and $\boldsymbol{\epsilon} = \boldsymbol{e}_i \cdot (\boldsymbol{e}_j \times \boldsymbol{e}_k)\,\boldsymbol{e}_i \otimes \boldsymbol{e}_j \otimes \boldsymbol{e}_k$ is the Levi-Civita permutation tensor. The set of normal vectors on the faces in the initial configuration is given by $\boldsymbol{n}_1^0 = \boldsymbol{e}_1$, $\boldsymbol{n}_2^0 = \frac{1}{2}(\boldsymbol{e}_1 + \sqrt{3}\boldsymbol{e}_2)$, $\boldsymbol{n}_3^0 = \frac{1}{2}(-\boldsymbol{e}_1 + \sqrt{3}\boldsymbol{e}_2)$, and $\boldsymbol{n}_4^0 = \boldsymbol{e}_3$. The first three vectors represent the three different orientations of the positive prism planes. The fourth vector describes the direction of the upper basal plane. Changing the signs of the vectors provides the opposing planes. The set $\mathcal{D} = \frac{1}{2}\{\sqrt{3}B, \sqrt{3}B, \sqrt{3}B, L\}$ assembles the distances of the prism planes to the grain origin $\boldsymbol{x}_0$ with the dimensionless breath B and length L.

The twelve vertices of the grains are given by the set

$$\mathcal{V} = \left\{\boldsymbol{v} \mid \boldsymbol{v} = \boldsymbol{x}_0 + B\,\boldsymbol{m}_i \pm \frac{1}{2}H\,\boldsymbol{n}_4,\ i = 1\ldots 6\right\}, \qquad (2.3)$$

with the edge vectors $\boldsymbol{m}_i = \frac{1}{2}(\boldsymbol{n}_{i\,\mathrm{mod}\,6+1}^{\mathcal{V}} + \boldsymbol{n}_i^{\mathcal{V}})$, $i = 1\ldots 6$, the normal vectors on the prism planes in positive and negative directions $\boldsymbol{n}^{\mathcal{V}} = \{\boldsymbol{Q}\boldsymbol{n}_i^0, -\boldsymbol{Q}\boldsymbol{n}_i^0\}$ and $i = 1\ldots 3$ and the axial vector $\boldsymbol{n}_4 = \boldsymbol{Q}\boldsymbol{n}_4^0$.

The set of 18 edges consists of three subsets. The set of six prism edges $\mathcal{E}^{\mathrm{P}}$ and two sets of each six basal edges $\mathcal{E}_1^{\mathrm{b}}$ and $\mathcal{E}_2^{\mathrm{b}}$ with

$$
\begin{aligned}
\mathcal{E}^{\mathrm{P}} &= \{\boldsymbol{x} \mid \boldsymbol{x} = \boldsymbol{v}_i + s(\boldsymbol{v}_{i+6} - \boldsymbol{v}_i)\} \\
\mathcal{E}_1^{\mathrm{b}} &= \{\boldsymbol{x} \mid \boldsymbol{x} = \boldsymbol{v}_i + s(\boldsymbol{v}_{i\,\mathrm{mod}\,6+1} - \boldsymbol{v}_i)\}, \qquad (2.4)\\
\mathcal{E}_2^{\mathrm{b}} &= \{\boldsymbol{x} \mid \boldsymbol{x} = \boldsymbol{v}_{i+6} + s(\boldsymbol{v}_{i\,\mathrm{mod}\,6+7} - \boldsymbol{v}_{i+6})\}
\end{aligned}
$$

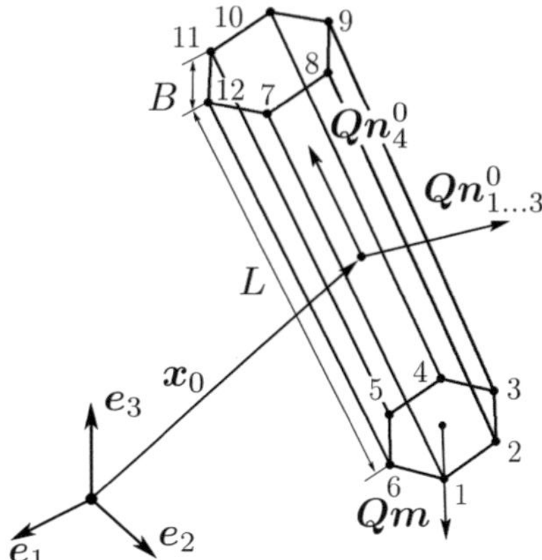

Figure 2.1: Grain lattice model in a general configuration: hexagonal prism with coordinate system, origin, prism normals, edge vectors and dimensions; vertice numbers correspond to the edge set $\mathcal{E}$ and vertice set $\mathcal{V}$.

where $s = [0, 1]$ and $i = 1 \ldots 6$. Therefore, the general set of edges can be combined into $\mathcal{E} = \mathcal{E}^{\mathrm{P}} \cup \mathcal{E}_1^{\mathrm{b}} \cup \mathcal{E}_2^{\mathrm{b}}$.

The eight faces $\mathcal{F}$ consist of six prism faces and two basal faces with the normal form

$$\mathcal{F} = \{x \mid x \cdot n_i^{\mathcal{F}} - n_i^{\mathcal{F}} \cdot (x_0 + \mathcal{D}_i n_i^{\mathcal{F}}) = 0, \ i = 1 \ldots 8\}. \tag{2.5}$$

The rotated outward normal vectors on the prism faces are $n_i^{\mathcal{F}} = \{Qn_i^0, -Qn_i^0\}$ with $i = 1 \ldots 4$.

The ℓ^p-norm is a generalization of the Euclidean norm ($p = 2$) with respect to both the exponent p and the used base vectors. The formal representation of the grains is given by the ℓ^p-norm on the projections $\mathcal{P}$ of the difference of a certain point x in space to the origin of the grain x_0 on the oriented normal planes with normal vector

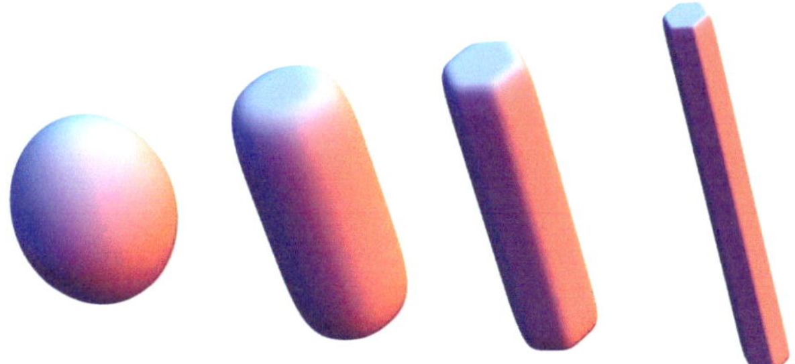

Figure 2.2: Contour of the anisotropic norm $\ell^p(\boldsymbol{x}) = 1$ with increasing exponents $p = \{2, 5, 10, 20\}$ and aspect ratios: The angularity of the grain models increases with increasing values for the exponent p. Natural shapes are obtained by high values, e.g., $p = 10\ldots20$ and $L/B = 3\ldots10$. The left contour is an ellipse, i.e., a Euclidean norm for $p = L/B = 2$.

$\boldsymbol{n}_i$ and the length scale $\mathcal{D}_i$

$$\mathcal{P}_i(\boldsymbol{x}) = \frac{1}{\mathcal{D}_i} \boldsymbol{Q}\boldsymbol{n}_i^0 \cdot (\boldsymbol{x} - \boldsymbol{x}_0), \quad i = 1\ldots4. \tag{2.6}$$

Hence, the norm can be written as

$$\ell^p(\boldsymbol{x}) = \sqrt[p]{\sum_{i=1}^{4} |\mathcal{P}_i(\boldsymbol{x})|^p}. \tag{2.7}$$

As limiting case $p \to \infty$, the ℓ^p-norm is transformed into the maximum norm

$$\ell^{\max}(\boldsymbol{x}) = \max\{|\mathcal{P}_i(\boldsymbol{x})|, \ i = 1\ldots4\}. \tag{2.8}$$

The anisotropy can be obtained by the choice of the normal vectors $\boldsymbol{n}$ and the dimensions $\mathcal{D}$. Figure 2.2 shows several cases. The transition from the Euclidean norm into an angular form with high aspect ratios can be observed.

2.2.3 Randomization

For a statistical structure generation a homogeneous statistical distribution of locations, orientations, and growth parameters are the precondition, which can deliver an isotropic orientation distribution, if enough grains are in the statistical ensemble. The origin of a grain x_0 depends on three random numbers $r_{1\ldots3}$ between 0 and 1. The vectors are scaled by the edge length w of the cube of consideration, such that the grain origin is $x_0 = w(r_1\,e_1 + r_2\,e_2 + r_3\,e_3)$. Here, it is useful to avoid points too close to each other to allow for a noteworthy growth. The penalty distance of a set of points is determined by the notion of an assumed mean density $\langle V^{\mathcal{G}} \rangle = w^3/n^{\mathcal{G}}$, with the intended number of grains in the unit cell $n^{\mathcal{G}}$. Thus, the mean distance is $\langle d^{\mathcal{G}} \rangle = \sqrt[3]{\langle V^{\mathcal{G}} \rangle}$, and the penalty distance is $d^{\mathrm{pen}} = k^{\mathrm{pen}}\langle d^{\mathcal{G}} \rangle$, with the penalty factor $0 < k^{\mathrm{pen}} < 1$. A new grain origin x_0^2, which is seeded next to an already existing grain origin x_0^1 is not used if $\|x_0^1 - x_0^2\| < d^{\mathrm{pen}}$.

Feasible values for the penalty factor k^{pen} are between $1/4$ and $1/2$. Figure 2.3 shows four examples in two dimensions. The impact of the penalty factor is significant. With increasing penalty factor, the mean distance between the objects increases due to the enforced distance around them. For the sequential adsorption undesirable grain clusters are avoided. These clusters would end in an early-stage growth hindrance, which results in structures with low aspect ratios and, therefore, unrealistic stereographic properties.

In case of reseeding, all new grain origins have to be checked for not being within the space of already existing grains. Here, the penalty distance $\Delta \mathcal{D}^{\mathrm{pen}}$ is used in the projection

$$\mathcal{P}_i^{\mathrm{pen}} = \frac{n_i^1 \cdot (x_0^2 - x_0^1)}{\mathcal{D}_i^1 + \Delta \mathcal{D}^{\mathrm{pen}}} \quad \Rightarrow \quad \ell_{12}^{\mathrm{pen}} = \max\{|\mathcal{P}_i^{\mathrm{pen}}|\}, \; i = 1\ldots4. \quad (2.9)$$

Reasonable values for $\Delta \mathcal{D}^{\mathrm{pen}}$ are in the range of $1\ldots5\%$ of the edge length w. The growth for the prismatic (breadth B) and basal direc-

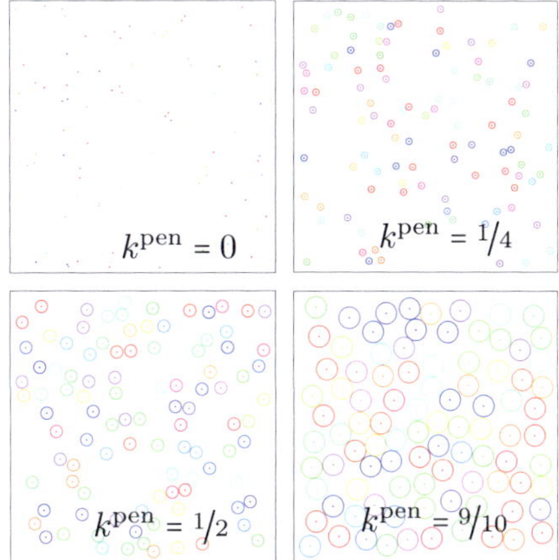

Figure 2.3: Influence of the penalty distance on the sequential adsorbed ensembles demonstrated on random seeded points in two dimensions with penalty factor $k^{\mathrm{pen}} = \{0, 1/4, 1/2, 9/10\}$. The magnitude of the penalty is indicated by the circles around the points. The higher the penalty factor, the larger the mean distance between neighboring objects.

tion (length L) relations are given by

$$
\begin{aligned}
k_B &= \left(1 + c_B(1 - 2r_4)\right) k_B^0, & \Delta B = k_B\,\Delta t, & \quad B = k_B\,t, & (2.10)\\
k_L &= \left(1 + c_L(1 - 2r_5)\right) k_L^0, & \Delta L = k_L\,\Delta t, & \quad L = k_L\,t, & (2.11)
\end{aligned}
$$

such that the growth velocities are scattering around $k_{B/L}(1 \pm c_{B/L})$. The homogeneous distribution of the random numbers on the unit sphere requires remapping of the givens $r_{6\ldots8}$ due to the curvature of the sphere. According to Shoemake (1992), the relations $\vartheta = \arccos(1 - 2r_6)$, $\varphi = 2\pi\,r_7$ and $\omega = \sin(\omega) - \pi\,r_8$ are used to achieve a homogeneous orientation distribution on the unit sphere $SO(3)$. The rotation angle ω has to be calculated numerically.

Hence, the whole system is determined by $8\,n^{\mathcal{G}}$ random numbers. These pseudo-random numbers are provided by the generator "sunif" (Ahrens and Kohrt, 1983).

2.2.4 Steric Hindrance

Experimental Observations

The development of the silicon nitride structure is strongly influenced by the steric impingement of the grains. Krämer et al. (1993) have made three important observations on pinning (Figure 2.4):

Observation 1: "The growth of a basal plane of a β-grain is stopped, when it hits a prism plane of another one as shown by arrow 1."

Observation 2: "If the thickness of a β-grain is comparatively large, its basal plane grows around a prism plane of a neighboring grain (arrow 2) through the 'free space' of liquid pockets and can include smaller grains."

Observation 3: "Edges and corners of prism planes are frequently rounded in case of edge-on-plane contact with other grains (arrow 3)."

Implementation of Krämer's First Observation: Growth and Pinning

The implementation of a conclusive growth-after-pinning behavior is one of the main achievements of this work. For the numerical implementation of the first observation an intersection of the 18 edges $\mathcal{E}_{\mathrm{s}}$ of a smaller grain (s) with the eight faces of the bigger grain (b), $\mathcal{F}_{\mathrm{b}}$ is considered. Here, the intersection points p^{int} are determined

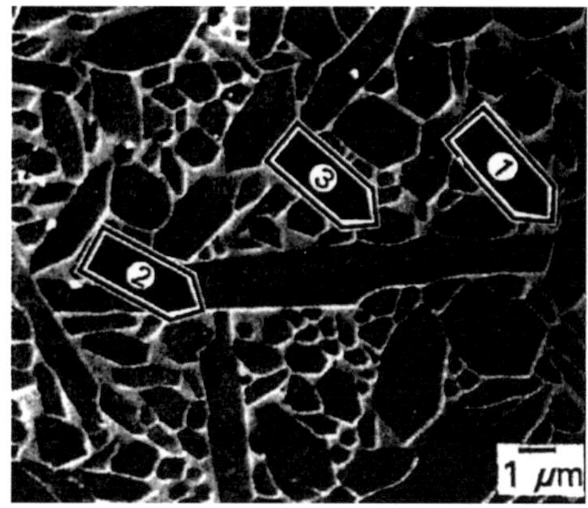

Figure 2.4: Three observations on grain pinning of Krämer et al. Krämer et al. (1993). Arrow 1 shows the case, when "a basal plane of a β-grain [...] hits a prism plane of another one [...]"; arrow 2 displays that when "the thickness of a β-grain is comparatively large its basal plane grows around a prism plane of a neighboring grain through the 'free space' of liquid pockets and can include smaller grains"; arrow 3 visualizes the rounding of grain boundaries "in case of edge-on-plane contact with other grains".

by $\mathcal{E}_s \cap \mathcal{F}_b$. The line segment for a certain edge is given by Eq. (2.4), such that $\boldsymbol{x}_s(s) \cdot \boldsymbol{n}_b = d_b$ is a projection of a point on an edge onto a face with the normal $\boldsymbol{n}_b$ and the distance of the face to the origin of the coordinate system $d_b = (\boldsymbol{x}_{0,b} + \mathcal{D}_b \boldsymbol{n}_b) \cdot \boldsymbol{n}_b$. The line segment parameter s is determined by $s = (d_b - \boldsymbol{v}_s^1 \cdot \boldsymbol{n}_b)/((\boldsymbol{v}_s^2 - \boldsymbol{v}_s^1) \cdot \boldsymbol{n}_b)$. When $\boldsymbol{x}^{\mathcal{V}_i^1}$ is a candidate for a valid intersection point $\boldsymbol{p}^{\text{int}}$, the parameter s is in its co-domain $[0,1]$. If this (necessary) condition is fulfilled, a second check is needed: The validity of $\boldsymbol{x}^{\mathcal{V}_i^1}$ can be seen if $\ell_b^{\max}(\boldsymbol{x}_s(s)) \leqslant 1$. If the sufficient condition is met as well, $\boldsymbol{p}^{\text{int}} := \boldsymbol{x}_s(s)$. In general, there will be at least three intersection points for a finite penetration of two grains. For the sake of simplicity, the mean value of these intersection points p^{m} is used to determine pinning case and post-pinning growth. After the determination of the mean intersection point p^{m}, the interaction case between the grains is determined from the projections

$$\mathcal{P}_i^*(p^{\mathrm{m}}) = \frac{n_i \cdot (p^{\mathrm{m}} - x_0)}{\mathcal{D}_i - \Delta\mathcal{D}}, \; i = 1\ldots 4. \tag{2.12}$$

The difference $\Delta\mathcal{D}$ has to be chosen because the mean-intersection points in general lie inside the grains due to the finite growth velocities in discrete pseudo-time increments. The mean length increment $k_L(1 + c_L)\Delta t$ is a feasible choice. For the determination of the pinning-cases, $\mathcal{P}_i^*(p^{\mathrm{m}}) = \pm 1$, $i = 1\ldots 4$ is evaluated. The contact with more than one grain and the related post-pinning growth can be described by averaging over the mean intersection points from all contacts, if both or none of the basal planes are involved in contact. For the remaining cases, a point in the region of the basal edges has to be used.

For the sake of brevity, the relevant information from the projections $\mathcal{P}^*$ is condensed into a boolean list $\mathcal{L}$ of length 8. The default value is $\mathcal{L}_i$ = false, $i = 1\ldots 8$ and means no contact. In case of contact, $\mathcal{L}$ is set true. The indices $1\ldots 3$ and $5\ldots 7$ designate the opposing prism faces. The indices 4 and 8 are for the upper and lower basal face.

$$\begin{aligned} \mathcal{L}_i &= \text{true} \quad \text{if} \quad \mathcal{P}_i^* = 1, \\ \mathcal{L}_{i+4} &= \text{true} \quad \text{if} \quad \mathcal{P}_i^* = -1 \end{aligned} \quad \text{and} \quad i = 1\ldots 4. \tag{2.13}$$

The following case consideration is used to describe the post-contact growth in the structure generator. It is a simple implementation of the first observation of Krämer et al. (1993). For cases not directly included reasonable approaches have been chosen, which avoid uncontrolled grain interpenetration and allow for high volume fractions and aspect ratios.

Case 1: No prism planes are in contact $\left(\wedge_{i=1}^{3}(\neg\mathcal{L}_i \wedge \neg\mathcal{L}_{i+4})\right)$:

- 1.1: No contact $(\neg\mathcal{L}_4 \wedge \neg\mathcal{L}_8)$: Default case, free growth in all directions.

- 1.2: One of the basal planes in contact $((\mathcal{L}_4 \vee \mathcal{L}_8) \wedge \neg(\mathcal{L}_4 \wedge \mathcal{L}_8))$: Free growth is allowed in all prism directions and axial growth is only possible in the opposite contact direction. The grain origin is shifted by $\Delta x = \mp 1/2 \, \Delta L \, n_4$.

- 1.3: Both basal planes in contact $(\mathcal{L}_4 \wedge \mathcal{L}_8)$. Grains can grow in all prism directions, however, no axial growth is allowed, which implies no shift of the grain origin.

Case 2: No opposing prism planes are in contact $\left(\vee_{i=1}^{3}[(\neg\mathcal{L}_i \wedge \mathcal{L}_{i+4}) \vee (\mathcal{L}_i \wedge \neg\mathcal{L}_{i+4})]\right)$:

- 2.1: No basal faces in contact $(\neg\mathcal{L}_4 \wedge \neg\mathcal{L}_8)$: Grains can grow further in both dimensions with half the velocity for the radial direction. The origin of the grain is shifted by $\Delta x_0 = 1/2 \, \Delta B \, s$ with the normalized shift vector s, which is determined by a projection of the difference vector on the mid plane in axial direction $s = (I - n_4 \otimes n_4)(x_0 - p^{\mathrm{m}})/\|x_0 - p^{\mathrm{m}}\|$.

- 2.2: One basal face in contact $((\mathcal{L}_4 \vee \mathcal{L}_8) \wedge \neg(\mathcal{L}_4 \wedge \mathcal{L}_8))$: Further growth is possible for both dimensions with half the velocity. The origin of the grain is shifted by $\Delta x_0 = 1/2 \sqrt{\Delta H^2 + 3\,\Delta B^2}\, s$ with the direction from the basal edge to the origin of the grain $s = (x_0 - p^{\mathrm{m}})/\|x_0 - p^{\mathrm{m}}\|$.

- 2.3: Both basal faces in contact $(\mathcal{L}_4 \wedge \mathcal{L}_8)$: The grain can only become thicker and the origin of the grain is shifted as described for case 2.1.

Case 3: Opposing prism planes in contact $\left(\bigvee_{i=1}^{3}(\mathcal{L}_i \wedge \mathcal{L}_{i+4})\right)$:

3.1: No basal faces in contact ($\neg\mathcal{L}_4 \wedge \neg\mathcal{L}_8$): Grain grows further only in axial direction with no shift of the grain origin.

3.2: One basal face in contact $((\mathcal{L}_4 \vee \mathcal{L}_8) \wedge \neg(\mathcal{L}_4 \wedge \mathcal{L}_8))$: Further axial growth is allowed only in opposite axial direction with half the velocity with origin shift as described for case 1.2.

3.3: Both basal faces in contact ($\mathcal{L}_4 \wedge \mathcal{L}_8$): Complete stop of growth.

Implementation of Krämer's Second Observation: Growing Around

The second observation of Krämer et al. (1993) is implemented by the consideration of the axis distance and the thickness of two grains 1 and 2. The distance between two non-parallel lines is given by

$$d_{12} = \min\{\|s_1 \boldsymbol{n}_4^1 - s_2 \boldsymbol{n}_4^2 - \boldsymbol{r}_{12}\|\} \tag{2.14}$$

with $\boldsymbol{n}_4^1 = \boldsymbol{Q}^1 \boldsymbol{n}_4^0$, $\boldsymbol{n}_4^2 = \boldsymbol{Q}^2 \boldsymbol{n}_4^0$ and $\boldsymbol{r}_{12} = \boldsymbol{x}_0^2 - \boldsymbol{x}_0^1$. The perpendicularity conditions $(s_1 \boldsymbol{n}_4^1 - s_2 \boldsymbol{n}_4^2 - \boldsymbol{r}_{12}) \cdot \boldsymbol{n}_4^1 = 0$ and $(s_1 \boldsymbol{n}_4^1 - s_2 \boldsymbol{n}_4^2 - \boldsymbol{r}_{12}) \cdot \boldsymbol{n}_4^2 = 0$, with $\boldsymbol{r}_{12} = \boldsymbol{x}_0^2 - \boldsymbol{x}_0^1$ yield the linear system

$$\begin{bmatrix} s_1 \\ s_2 \end{bmatrix} = \frac{1}{1 - (\boldsymbol{n}_4^1 \cdot \boldsymbol{n}_4^2)^2} \begin{bmatrix} 1 & -\boldsymbol{n}_4^1 \cdot \boldsymbol{n}_4^2 \\ \boldsymbol{n}_4^1 \cdot \boldsymbol{n}_4^2 & -1 \end{bmatrix} \begin{bmatrix} \boldsymbol{r}_{12} \cdot \boldsymbol{n}_4^1 \\ \boldsymbol{r}_{12} \cdot \boldsymbol{n}_4^2 \end{bmatrix}. \tag{2.15}$$

It is obvious that this approach is not feasible for the special case of parallel axes. So, the projection of the vector $\boldsymbol{r}_{12}$ on the perpendicular vectors is used. Then the distance is

$$d_{12} = \|(\boldsymbol{I} - \boldsymbol{n}_4^1 \otimes \boldsymbol{n}_4^1)\boldsymbol{r}_{12}\| = \sqrt{\boldsymbol{r}_{12} \cdot \boldsymbol{r}_{12} - (\boldsymbol{r}_{12} \cdot \boldsymbol{n}_4^1)^2}. \tag{2.16}$$

Hence, the condition for "growing around" is $d_{12} > k^{\mathrm{ga}}(B_1 + B_2)$ with $0 < k^{\mathrm{ga}} < 1$ as the allowed ratio of interpenetration. If this condition is fulfilled there will be no further examination in the recent time step.

Implementation of Krämer's Third Observation: Grain Overlapping and Rounding of Grain Interfaces

The third observation of Krämer et al. (1993) is implemented in the context of the voxel discretization, which is used as an input for the finite element mesh generation. Here, a certain overlap of grains is reached by an offset value $\Delta\mathcal{D}^{\mathrm{o}}$ to the exact dimension $\mathcal{D}$ similarly to the offset, which allows for the usage of the projection information. The ℓ^p-norm for the pixel generation then takes the form

$$\ell_{\mathrm{o}}^{p}(\boldsymbol{x}) = \sqrt[p]{\sum_{i=1}^{4}\left|\frac{\boldsymbol{n}_i \cdot (\boldsymbol{x} - \boldsymbol{x}_0)}{\mathcal{D}_i + \Delta\mathcal{D}^{\mathrm{o}}}\right|^{p}}. \tag{2.17}$$

The constant $\Delta\mathcal{D}^{\mathrm{o}}$ is used for the scaling of length and breadth of the grains. So, the ratio of length to breadth is not preserved. The shape of the overlap regions is obtained by the consideration of ℓ_{o}^{p} for two grains 1 and 2: $\ell_{\mathrm{o},1}^{p}(\boldsymbol{x}) \leqslant 1 \wedge \ell_{\mathrm{o},2}^{p}(\boldsymbol{x}) \leqslant 1$, which means that a certain point $\boldsymbol{x}$ belongs to both grains. An additional criterion has to be used to decide to which grain the point $\boldsymbol{x}$ belongs. A natural choice is the usage of the smaller norm as an indicator for the affiliation, where $\boldsymbol{x}$ is assumed to belong to the grain with the smaller ℓ_{o}^{p}. The exponent p is used to define the angularity of the grains and with it the roundness of the overlapping region. Figure 2.5 shows three cases for overlapping. The effect of overlapping and rounding of the boundary significantly depends on the exponent p.

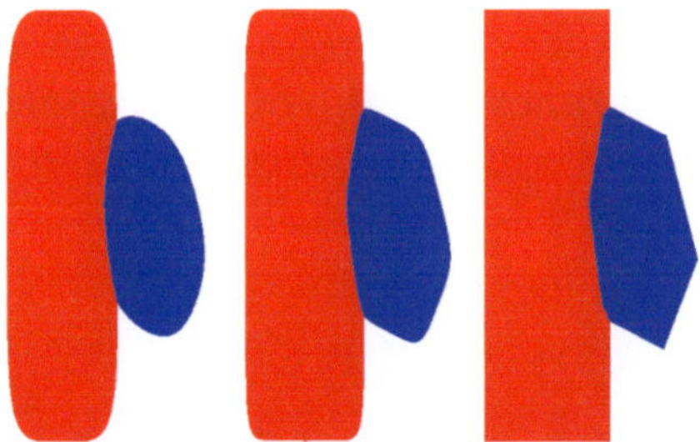

Figure 2.5: Demonstration of the implementation of observation 3 of Krämer et al. (1993): overlapping of grains for a fixed overlapping distance $\Delta\mathcal{D}^{\circ}$ = $1/2\,B$ and increasing norm exponents $p = \{5, 10, \infty\}$ (from left to right). The angularity of the overlapping zone increases with the norm exponent. Naturally rounded overlapping and grain shapes can be obtained by a norm exponent of $p = 10$.

2.2.5 Periodization

The periodicity of the created structures is desirable for two reasons. First, for the consideration of a mechanical problem, the boundary conditions play a crucial role. It is well-known that periodic displacement boundary conditions together with periodic structures are a prerequisite for the determination of a realistic effective material behavior, see, e.g., Suquet (1982). Second, the consideration of geometric properties of a periodic structure is much more straight forward than for non-periodic ensembles where boundary effects like intersected grains are inevitable.

The periodization of the generated microstructure is obtained by the consideration of 26 periodization boxes, in which periodically shifted copies of the seeded grains are interacting. The origin x_0 is shifted by a periodization vector

$$\Delta \boldsymbol{x}^{\mathrm{P}}_{ijk} = w(i\,\boldsymbol{e}_1 + j\,\boldsymbol{e}_2 + k\,\boldsymbol{e}_3), \ \boldsymbol{x}^{\mathrm{P}}_{ijk} = \boldsymbol{x}_0 + \Delta \boldsymbol{x}^{\mathrm{P}}_{ijk}, \ ijk = -1, 0, 1. \quad (2.18)$$

For every periodic copy of a grain, the considerations in Section 2.2.4 have to be carried out. Additionally, the mean intersection points have to be kept up to date. This means that for every interaction case of two grain copies the recent intersection points $\boldsymbol{p}^{\mathrm{m, rec}}_{ijk,n}$, $n = 1, 2$ have to be updated. All other intersection points $\boldsymbol{p}^{\mathrm{m}}_{ijk,n}$ have to be updated as well, which is done by

$$\boldsymbol{p}^{\mathrm{m}}_{ijk,n} = \boldsymbol{p}^{\mathrm{m, rec}}_{ijk,n} - \Delta \boldsymbol{x}^{\mathrm{P, rec}}_{ijk,n} + \Delta \boldsymbol{x}^{\mathrm{P}}_{ijk}, \ n = 1, 2, \quad (2.19)$$

where $\Delta \boldsymbol{x}^{\mathrm{P, rec}}_{ijk,n}$ is the offset of the recent grain copy.

2.3 Results

2.3.1 Pseudo-Time Evolution of the Microstructure

The pseudo-time evolution of statistical geometric quantities will be discussed in this section. The best known property of microstructures is the volume fraction of particles. As pointed out by, e.g., Becher et al. (1998), the distribution of grain size and aspect ratio plays a crucial role for the fracture toughness.

Thus, Figure 2.6 assembles the pseudo-time evolution of the volume fraction, the mean aspect ratio and its standard deviation. The volume of one grain is determined by $V = 3/2\sqrt{3}B^2 H$. The accumulated grain volume is $V^{\mathcal{G}}_{\mathrm{acc}} = \sum_{i=1}^{n^{\mathcal{G}}} V_i$. The grain volume fraction is the ratio of the accumulated grain volume to the volume of the region of interest $c^{\mathcal{G}}_{\mathrm{acc}} = V^{\mathcal{G}}_{\mathrm{acc}}/w^3$. The volume fraction of a single grain normalized on the accumulated grain volume is $c = V/V^{\mathcal{G}}_{\mathrm{acc}}$. The aspect ratio $\mathcal{A}$ of a grain is the ratio of its length to its breadth.

For hexagonal prisms $\mathcal{A} = H/\sqrt{3}B$, if the width across flats is regarded as the small diameter. For the determination of the mean value of $\mathcal{A}$, the grain volume $V_{\text{acc}}^{\mathcal{G}}$ is used as the basis volume, such that $\langle \mathcal{A} \rangle = \sum_{i=1}^{n^{\mathcal{G}}} c_i \mathcal{A}_i$, where the grain volume fractions c_i, $i = 1 \ldots n^{\mathcal{G}}$ act as weights of the sum. The standard deviation is calculated by

$$\sigma(\mathcal{A}) = \sqrt{\sum_{i=1}^{n^{\mathcal{G}}} c_i (\mathcal{A}_i - \langle \mathcal{A} \rangle)^2}. \tag{2.20}$$

The following figures show the evolution of a periodic ensemble of grains in a cell of width $w = 16$ without any overlaps after the second case. It is, therefore, an exact geometric ensemble consisting of non penetrating hexagons. The generation was carried out in 105 seed steps of each 10 grains. So, a total of $27 \times 105 \times 10 = 28{,}350$ entities have to be considered in the final seed step. The growth velocity parameters in Eqs. (2.10) and (2.11) are $k_L = 10 k_B = 10$, the penalty factors after Section 2.2.3 and Eq. (2.9) are $k^{\text{pen}} = 1/10$ and $\Delta \mathcal{D}^{\text{pen}} = 1/10$.

Figure 2.6 shows the pseudo-time evolution of the accumulated grain volume fraction $c_{\text{acc}}^{\mathcal{G}}$. The steep accession in the beginning is due to the free growth of the grains, mainly without steric hindrance. At a grain volume fraction of around 25% the grain interaction gains increasing influence, which causes a certain saturation of the grain volume fraction. The final value of $c_{\text{acc}}^{\mathcal{G}}$ lies at 50%, which can be considered to be a high value for a complicated randomized structure without overlapping (Cooper, 1988). The trend at the end of the process goes already towards increasing volume fraction.

In Figure 2.7 the mean aspect ratio $\langle \mathcal{A} \rangle$ with its scatter band $\pm \sigma(\mathcal{A})$ is depicted. Here, it can be seen that a structure with aspect ratios of approximately 6 and a relatively wide variance of approximately 2 is generated using the mentioned generator setup. The mean aspect ratio of the structures reaches a local peak at the pseudo-time, when the grains come into interaction, i.e., the pseudo-time, when the

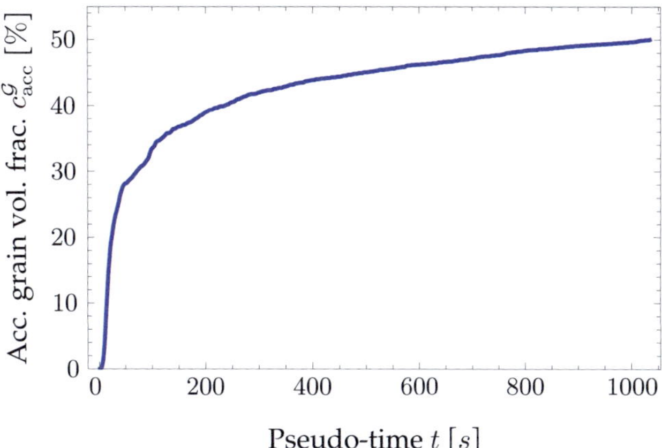

Figure 2.6: Pseudo-time evolution of the accumulated volume fraction $c_{acc}^{\mathcal{G}}$ of a periodic 105×10 grain ensemble; after a steep increase in the beginning of the adsorption-and-growth process, saturation takes place, due to the steric hindrance.

saturation of the volume fraction begins. Beyond this point in time, the aspect ratio is monotonically decreasing, because the hindrance in axial direction is more likely than in radial direction.

It is important to gain insight into the influence of the parameters for the control of the growth process. Therefore, parameter variations on the base of the already mentioned set of input values have been carried out. The following diagrams show the mean aspect ratio $\langle A \rangle$ over the grain volume fraction $c_{acc}^{\mathcal{G}}$, due to their significance for the silicon nitride microstructure.

One important factor is the growth velocity parameter k_L. It has been varied, since it has a noteworthy impact on the aspect ratio of the obtained structures. In the beginning of the process, a broad distribution can be observed. The seeding penalty factor $k^{pen} = 1/2$ is chosen to enforce more space for the structure generation pro-

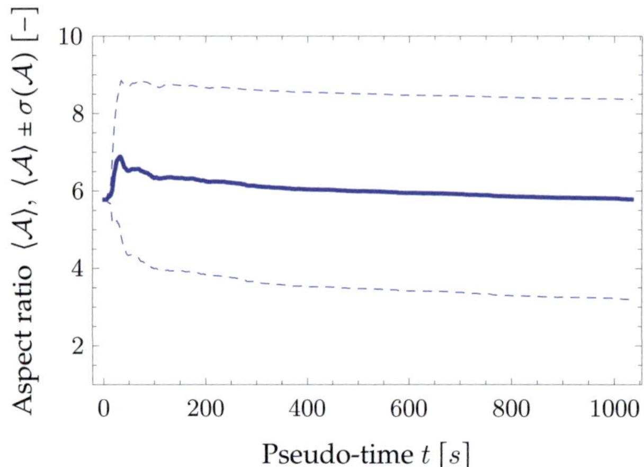

Figure 2.7: Pseudo-time evolution of the mean aspect ratio $\langle \mathcal{A} \rangle$ (—) with scatter band $\langle \mathcal{A} \rangle \pm \sigma(\mathcal{A})$ (--) for a periodic 105×10 grain ensemble; after a peak at the beginning of the structure generation process, the mean value decreases slowly, the scattering remains at a constant level.

cess with higher growth anisotropies (Section 2.2.3). After a peak of the aspect ratio, which is shifted to lower volume fractions for greater axial growth velocities, all curves appear converging at a mean aspect ratio of approximately 6, and a volume fraction of approximately 60%. So, merely a temporary impact on the mean value of the aspect ratio is observed by this factor. The relevance for the distribution $\mathcal{A}/L$ will be shown in Section 2.3.2.

2.3.2   Distribution of the Geometric Quantities

The statistical distribution of geometric quantities is important for a comparison of artificial microstructures with experimental data. Additionally, it is relevant to gain further insight into the behavior

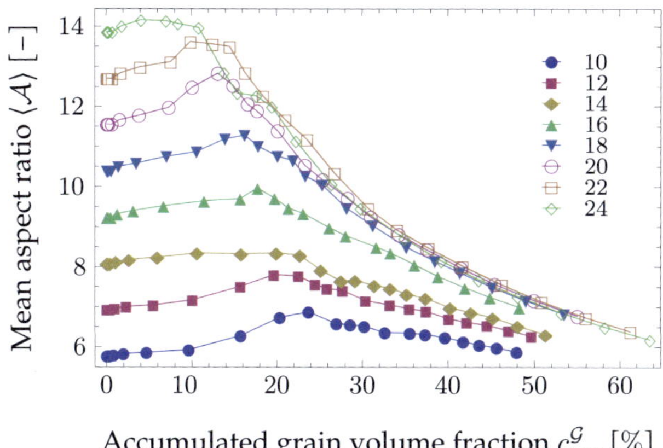

Figure 2.8: Mean aspect ratio $\langle \mathcal{A} \rangle$ over grain volume fraction $c^{\mathcal{G}}_{\mathrm{acc}}$ of a periodic 105×10 grain ensemble for a variation of the ratio of the growth constants $k_L/k_B = 10 \ldots 24$ and $k_B = 1 = \mathrm{const.}$, $k^{\mathrm{ga}} = 9/10$, $c_{B/L} = 1/5$, $k^{\mathrm{pen}} = 1/2$, $\Delta \mathcal{D}^{\mathrm{pen}} = 1/10$; The impact of the axial grain growth velocity on the mean aspect ratio is significant only in the beginning of the structure generation process.

of the structural model. So, the example of Section 2.3.1, Figure 2.8 is readopted. Figure 2.9 depicts probability densities of the aspect ratio and the grain length for different grain growth anisotropies. The probability density is the grain volume fraction normalized on the total grain volume. In order to adjust the grain volume fraction to from 60% to the natural fraction of 88%, an overlapping factor of $\Delta \mathcal{D}^{\circ} = 0.262$ has been chosen, see Eq. (2.17) and Figure 2.5. Natural slightly rounded grain edges have been obtained by a norm exponent $p = 20$.

The influence of the growth velocity anisotropy can, now, be seen in detail. Although similar mean aspect ratios (and grain lengths) are observed in the pseudo-time evolution in Figure 2.8, the distribution is obviously different. A trend for higher aspect ratios and longer grains with leaner scattering is observed for increasing growth anisotropy.

The experimental reference data has been acquired from SEM micrographs of the commercial silicon nitride grade SL 200 BG by an elaborated digital image processing technique (Fünfschilling et al., 2011). The main problem in measuring structural data of complex microstructures is the fact that an elongated grain can be intersected in many different ways, see, e.g., Ohser and Mücklich (2000). A rational estimation of the grain length and aspect ratio is, therefore, in general a contingency. Figure 2.10 shows the problem: The grain is intersected by three different planes. The brown plane (•) contains the grain axis. Hence, the grain's length and aspect ratio is represented realistically. Differently the green plane (•): The whole cross section is intersected, but in a skew angle, such that the grain appears much smaller and with significantly decreased aspect ratio, than it has in reality. The limiting case would be a section plane, which is perpendicular to the grain axis. Here, any information on grain length and aspect ratio would be lost. A further undesirable but also inevitable case is represented by the blue plane (•), which is intersecting only a small part of the whole cross section. The result is an irregular polygon, which does not even allow for a serious prediction of the grain thickness.

A well established procedure is the transformation of measurable two-dimensional distributions of size and shape into distributions of spatial size and shape. Mücklich et al. (1994) applied this technique first to the demanding case of silicon nitride. The idea is that a planar section through a spatial distribution of (non-overlapping) hexagonal prisms will result in a distribution of convex intersection polygons. So, a stereological function in form of a linear map can be formulated as link between planar and spatial distribution of size and shape quantities. The kernel of this function for complicated shapes like hexagonal prisms can only be calculated in a

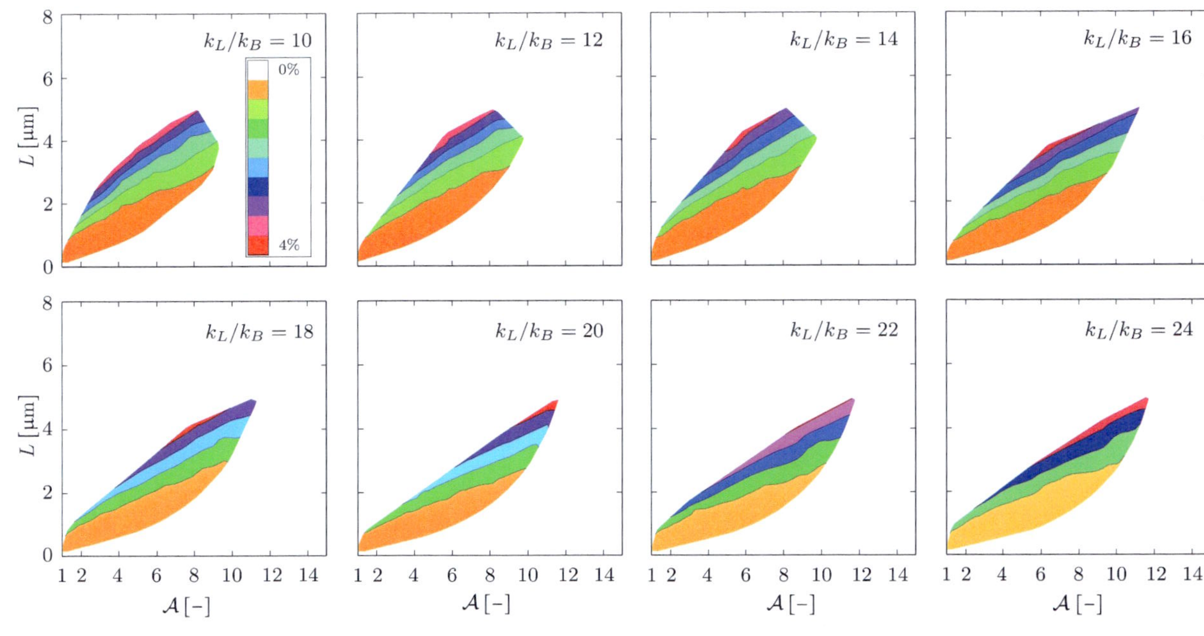

Figure 2.9: Cumulative frequency of aspect ratios of a periodic 105×10 grain ensembles for the parameter setup $k_L/k_B = 10\dots24$, $k^{\mathrm{ga}} = 9/10$, $c_{B/L} = 1/5$, $k^{\mathrm{pen}} = 1/2$, $\Delta\mathcal{D}^{\mathrm{pen}} = 1/10$, $\Delta\mathcal{D}^{\mathrm{o}} = 0.262$ and $p = 20$

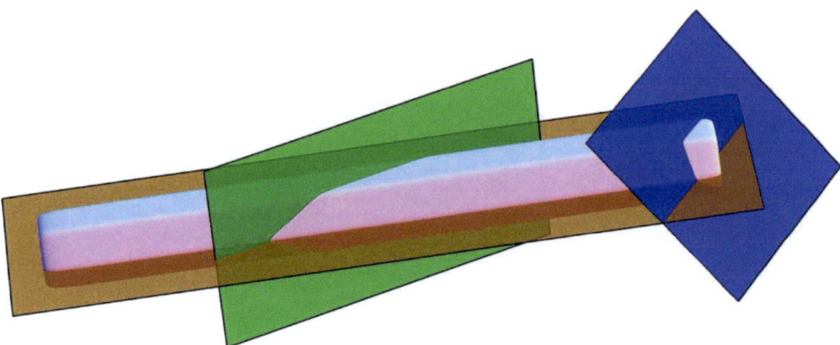

Figure 2.10: Grain with different section planes: An axis parallel section (•) delivers the real representation of the grain shape. A section through the grain axis (•) results an underestimation of the real aspect ratio. The edge section (•) figure contains completely unrealistic information about the grain shape.

discretized way due to the huge amount of intersection possibilities (Figure 2.10). In this line, simulations with randomly intersected hexagonal prisms have been carried out in order to acquire the coefficients of the kernel function. The relative frequencies of the spatial quantities then can be calculated by the EM algorithm, see Vardi et al. (1985), which was introduced into stereology by Silverman et al. (1990).

Fünfschilling et al. (2011) measured the microstructural properties of the commercial silicon nitride grade SL 200 BG with the image processing software ImageC (Aquinto, Berlin, Germany) and the stereology extension based on the work of Mücklich et al. (1994). For the measurement, an ensemble of approximately 3,000 grains has been examined.

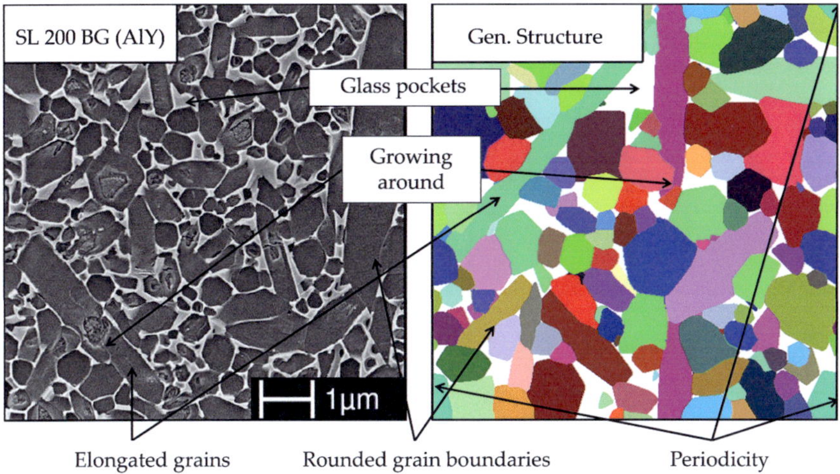

Figure 2.11: SEM-image of the AlY-doped commercial silicon nitride grade SL 200 BG Fünfschilling et al. (2011) and an exemplary section image of the generated periodic three-dimensional microstructure with 105×10 grains and the parameters $k_L/k_B = 22$, $k^{\mathrm{ga}} = 9/10$, $c_{B/L} = 1/5$, $k^{\mathrm{pen}} = 1/2$, $\Delta\mathcal{D}^{\mathrm{pen}} = 1/10$, $\Delta\mathcal{D}^{\mathrm{o}} = 0.262$ and $p = 20$.

The generated microstructure is determined by the grain growth velocity parameter combination $k_L/k_B = 22$. The other parameters are as described in Section 2.3.1 and in Figures 2.8 and 2.9.

Figure 2.11 juxtaposes a SEM micrograph of SL 200 BG and a two-dimensional section through the artificial three-dimensional structure with the mentioned parameters. In both cases, large and elongated as well as smaller and more roundish grains can be seen. A certain similarity between the two structures cannot be neglected. Therefore, a view to the most important geometrical properties of grain length and aspect ratio delivers further insight.

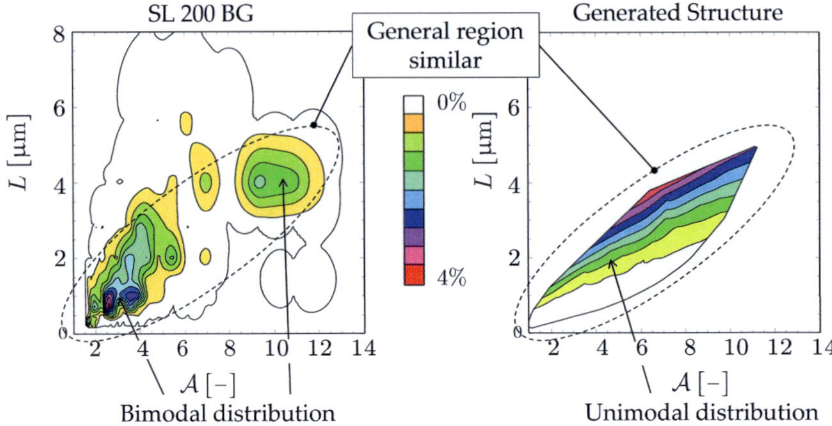

Figure 2.12: Cumulative frequency of aspect ratios in SL 200 BG Fünfschilling et al. (2011) and in a periodic 105×10 grain ensemble for $k_L/k_B = 22$, $k^{\mathrm{ga}} = 9/10$, $c_{B/L} = 1/5$, $k^{\mathrm{pen}} = 1/2$, $\Delta\mathcal{D}^{\mathrm{pen}} = 1/10$, $\Delta\mathcal{D}^{\circ} = 0.262$ and $p = 20$. The distributions of the experimental data and the generated structure are mainly in the same region, however the bimodality of the data is not captured by the simulation.

Figure 2.12 shows two diagrams with the probability density as function of aspect ratio and grain length. For SL 200 BG, the probability density is calculated after Mücklich et al. (1994), as described before. The diagram for the generated structure uses the grain volume fraction of the single grains in the three-dimensional structure as probability density. The two diagrams show certain differences: The SL 200 BG plot has several peaks. Two main peaks are clearly recognizable. The first one is at small grain lengths beneath 1 μm and has local peaks at aspect ratios between 2 and 4. These peaks represent grain shape/grain length combinations with minor relevance for the fracture toughness of the material, because the grains are too small to bear the high local loads around crack tips (Ohji et al., 1995; Becher et al., 1998). The other peak is at high grain lengths of 4 μm and high aspect ratios of approximately 10. This peak has a close relationship for the high fracture toughness of this material, because it means that the material contains a significant amount of strong and elongated grains, which can act as bridging grains and, therefore, avoid quick crack propagation (Fünfschilling et al., 2011). Interesting is the observation that a region between those peaks exists, in which hardly any grain can be found.

The diagram for the generated microstructure covers a similar domain, although it shows differences as well. The distribution does not show strongly pronounced peaks and its global maximum is not at the same position as in the stereological observation. Nevertheless, the main trend is towards long and elongated grains, which appear at similar grain lengths and aspect ratios. Reasons for the differences between the two diagrams can be attributed to a simplification of the structure evolution in the presented algorithm and to the differences with respect to the acquisition of the underlying data.

The comparison of the results from the thermodynamical simulations of Kitayama et al. (1998a) with the data, which was gained from the structure generation process shows good agreement, as well. Although, it has to be mentioned that the distributions of grain length over grain width and aspect ratio over grain width from the presented algorithm are broader with the chosen parameter combinations.

2.4   Summary and Conclusions

A simplified algorithm for the generation of periodic three-dimensional silicon nitride-like structures has been introduced. Its primal result is the creation of microstructural models for micromechanical finite element simulations. The microstructure generator is based on a sequential adsorption procedure, which is enriched with growth of particles, steric hindrance and overlapping, motivated by experimental observations of Krämer et al. (1994). These observations have been implemented in algorithmic detail to preserve high aspect ratios and grain volume fractions in the artificial microstructures.

In order to give an insight into the behavior of the model, the pseudo-time evolution of mean aspect ratio and grain volume fraction have been discussed for different parameter sets of a periodic ensemble of 1050 grains, which have been adsorped in 105 steps. It was shown that the mean aspect ratio in generated microstructures decreases after a temporary peak at approximately 20% grain volume fraction due to steric hindrance.

The impact of the axial growth rate on the pseudo-time evolution and the distribution of the geometric quantities aspect ratio and grain length has been examined. The pseudo-time evolution showed a certain convergence of aspect ratios at values of approximately 6 and volume fractions ratios of approximately 60%, which could be obtained without overlapping. Realistic grain volume fractions have been adjusted by variation of the grain overlapping. The adjusted ensembles have been considered with respect to the probability density of the aspect ratio and grain length. A clear redistribution of the densities to longer grains and with higher aspect ratios was caused by increased grain growth anisotropy.

The significance of the model has been shown by a comparison of the grain length and the aspect ratio distribution with literature values (Fünfschilling et al., 2011), showing reasonable correspondence. Hereby, a general accordance of the stereological SEM image evaluation with the data provided by the structure generator was observed. However, differences have been observed as well, concerning the bimodality of the probability density of grain length and aspect ratio. On the one hand, these can be attributed to the assumptions, which have been made for the implementation of the structural model. On the other hand, the procedures of the acquisition of the distributions for the SEM images and for the data from the structure generator is different, which should induce, as well, differences in the obtained distributions.

Aside from these issues, the structure generator reproduced the main features of a real silicon nitride microstructure, which has been shown by a comparison of a SEM micrograph of SL 200 BG and a synthetic microstructure.

Based on the aforementioned results, it can be concluded that the algorithmic structural model can, in general, be applied for generating realistic silicon nitride microstructures. Considering all discussed features, we conclude that the model appears as a reasonable approximation for the complex reality of the silicon nitride structure. Thus, it can be used for the numerical considerations of complicated material behaviors like thermoelasticity and trans- as well as intergranular fracture, which will be presented in the following chapters.

The general technique of sequential adsorption with growth and steric hindrance of particles is applicable to many different types of materials.

Chapter 3

Thermoelastic Properties[*]

3.1   Introduction

Silicon nitride is a structural reinforced ceramic, which is used for high performance applications due to its fair compromise of stiffness, strength, and toughness. An important feature is the excellent thermal shock resistance due to the low thermal expansion coefficient and a high thermal conductivity.

This interesting property profile is related to different aspects of the material on the microlevel. The elastic stiffness and the thermal expansion are dominated by the rod-like β-crystals, which are occupying the largest part of the material volume. Due to the heterogeneity of the microstructure (Chapter 2) and the phase contrasts between the grains and the glassy matrix thermal residual stresses on the microlevel are caused, when the material is exposed to changing temperatures. This chapter aims towards a description of these effects on the microlevel and an evaluation of the results in

---

[*]This chapter is based on the paper "Homogenization of the Thermoelastic Properties of Silicon Nitride" (Wippler et al., 2011).

the calculation of the effective thermoelastic properties. The results are applied in the fracture simulations in Chapter 4.

The rod-like single crystals have transverse-isotropic thermoelastic material properties. Henderson and Taylor (1975) presented temperature-dependent data for the thermal expansion between room temperature and 1020°C by X-ray diffraction methods. Vogelgesang et al. (2000) have determined the five transverse-isotropic elastic constants of β-Si_3N_4 grains by Brillouin scattering at room temperature.

Peterson and Tien (1995) determined the effects of residual stresses by experiments and Eshelby's inclusion method. Their finding is that residual stresses in the grain boundary phase due to thermal expansion increase the number of bridging grains and thereby, the fracture toughness. A visible result of this effect are the complex fracture patterns in the strongly heterogeneous microstructures.

In particular, the determination of thermal residual stresses in silicon nitride due to cooling down from the glass transition temperature after sintering on the microscale and the calculation of effective thermoelastic properties based on experimental values on the microscale for thermal expansion coefficients and elastic stiffnesses will be addressed in the chapter.

An important problem is that all elastic properties are temperature dependent. Additionally, not all properties on the microlevel are known. Due to the potential complexity of a meaningful analytical or semi-analytical approach, a numerical approach has been chosen. The decrease of stiffness with temperature in the grains could not be obtained from literature. So, it was chosen to be determined by an inverse consideration: If the effective and all microscopic parameters, except the temperature dependence of the grain stiffness are known, it can be obtained by an adjustment of the numerically predicted effective material behavior to the effective experimental data.

The uniqueness of this inverse procedure is preserved, because the temperature dependence of the microscopic grain stiffness is assumed to be determined by an isotropic scaling with a single constant. This choice is made according to the experimental observations on the effective Young's modulus of silicon nitride by Lube and Dusza (2007).

The chapter presents new experimental results on the thermal expansion behavior of different silicon nitride compositions. The modeling of the thermoelastic behavior of β-Si_3N_4 and of the glassy phase is described in detail. A novel approach for the implementation of periodic boundary conditions is presented and the determination of the effective material parameters is explained. The numerical results for the effective temperature dependent elastic stiffness tensor, the thermal strains and thermal expansion coefficient are comprised, and, in order to estimate the quality of the calculations, information on the isotropy of the effective tensors of stiffness and thermal expansion will be provided. The local loading within the discretized microstructures has been evaluated by an examination of the maximum principal stress, the hydrostatic pressure, the strain energy density and different measures of stress triaxiality. A considerable agreement with the Eshelby based approach pursued by Peterson and Tien (1995) was found for the residual stresses.

3.2 Measurement of the Thermal Expansion

Materials for the determination of the thermal expansion coefficients were produced by a two step sinter/hot isostatic pressing (HIP) process.[†] The sample denoted as MgLu had 8.5 wt% Lu_2O_3 and 1.93 wt% MgO as sintering aids. The composition MgY has 5 wt% Y_2O_3 and 2 wt% MgO. SL 200 BG is a commercial sili-

---

[†]The experiments have been carried out at the Institute for Ceramics in Mechanical Engineering of the KIT by Stefan Fünfschilling.

con nitride grade with Y_2O_3 and Al_2O_3 as sintering aids (CeramTec, Plochingen, Germany, (Lube and Dusza, 2007)). The thermal expansion of the three silicon nitride ceramics has been measured in a differential dilatometer (Baehr Model 402, Baehr Thermoanalyse GmbH, Hüllhorst, Germany) equipped with an alumina tube and push rods. The sample geometry was $3\times4\times10$ mm^3. The end faces were ground plane-parallel. The temperature cycle applied to the sample included heating from room temperature (20°C) at a rate of 2 K/min to 1000°C and subsequently cooling back to the room temperature at the same rate. The heating as well as the cooling part was used to calculate the thermal expansion coefficient. Temperature was monitored by a type S (Pt10%Rh-Pt) thermocouple. Based on a reference measurement of a sapphire standard, a correction function of the measurement data was calculated and applied to the measurement data according to the Baehr standard software procedure. The experimental results are shown together with the results from the numerical homogenization in Figure 3.8.

3.3 Material Model

3.3.1 Thermoelastic Model for β-Si$_3$N$_4$

Stiffness Tensor. The constituents of silicon nitride, i.e., the rod-like β-Si$_3$N$_4$ grains and the glassy phase formed by the sintering additives are expected to show a clear temperature dependence of both the elastic stiffness and the thermal expansion coefficients. The thermal expansion of the rod-like grains is provided by Henderson and Taylor (1975). Hampshire et al. (1994) examined the temperature dependencies of the elastic stiffness and the thermal expansion of oxynitride glasses, which are assumed as a model material for the glassy phase in silicon nitride. The thermoelastic form of

Hooke's law,

$$\boldsymbol{\sigma}(\boldsymbol{\varepsilon}, T) = \mathbb{C}(T)[\boldsymbol{\varepsilon} - \boldsymbol{\varepsilon}_{\text{th}}(T)], \tag{3.1}$$

is adopted in a geometrically linear setting for both phases. The Cauchy stress tensor and the infinitesimal strain tensor are represented by $\boldsymbol{\sigma}$ and $\boldsymbol{\varepsilon}$, respectively. Both, the stiffness tensor $\mathbb{C}(T)$ and thermal strain tensor $\boldsymbol{\varepsilon}_{\text{th}}(T)$ are assumed to be temperature dependent.

Due to the lack of experimental data, the stiffness tensor $\mathbb{C}^{\mathcal{G}}(T)$ of the rod-like grains $(\mathcal{G})$ is assumed to decrease linearly with rising temperature. It is assumed to be proportional to the stiffness $\mathbb{C}^{\mathcal{G}}(T_0)$ at the room temperature T_0. As a result of this simplifying assumptions, the temperature dependent stiffness tensor of the grains can be written as

$$\mathbb{C}^{\mathcal{G}}(T) = (1 + k^{\mathcal{G}} \Delta T) \mathbb{C}^{\mathcal{G}}(T_0) \tag{3.2}$$

with the temperature difference $\Delta T = T - T_0$. The single constant $k^{\mathcal{G}} < 0$ in Eq. (3.2) is used to adjust the thermoelastic model to experimental data of the macroscopic thermoelastic behavior. The rod-like grains show a transverse-isotropic material symmetry. Hence, the elastic stiffness tensor has the following form

$$\mathbb{C}^{\mathcal{G}}(T) = C^{\mathcal{G}}_{\alpha\beta}(T) \boldsymbol{B}_\alpha \otimes \boldsymbol{B}_\beta, \tag{3.3}$$

where the symmetric orthonormal basis tensors of second-order $\boldsymbol{B}_\alpha$ have been defined by Federov (1968): The normal directions are represented by $\boldsymbol{B}_1 = \boldsymbol{e}_1 \otimes \boldsymbol{e}_1$, $\boldsymbol{B}_2 = \boldsymbol{e}_2 \otimes \boldsymbol{e}_2$ and $\boldsymbol{B}_3 = \boldsymbol{e}_3 \otimes \boldsymbol{e}_3$. The shear directions are given by $\boldsymbol{B}_4 = \sqrt{2}\,\text{sym}(\boldsymbol{e}_1 \otimes \boldsymbol{e}_2)$, $\boldsymbol{B}_5 = \sqrt{2}\,\text{sym}(\boldsymbol{e}_2 \otimes \boldsymbol{e}_3)$ and $\boldsymbol{B}_6 = \sqrt{2}\,\text{sym}(\boldsymbol{e}_3 \otimes \boldsymbol{e}_1)$. The symmetry is preserved by the operator $\text{sym}(\boldsymbol{A}) = 1/2\,(\boldsymbol{A} + \boldsymbol{A}^{\top})$. The axial direction and with it the basal planes are defined by $\boldsymbol{e}_3$. For the orientation of the basis tensors and with them of the stiffness and

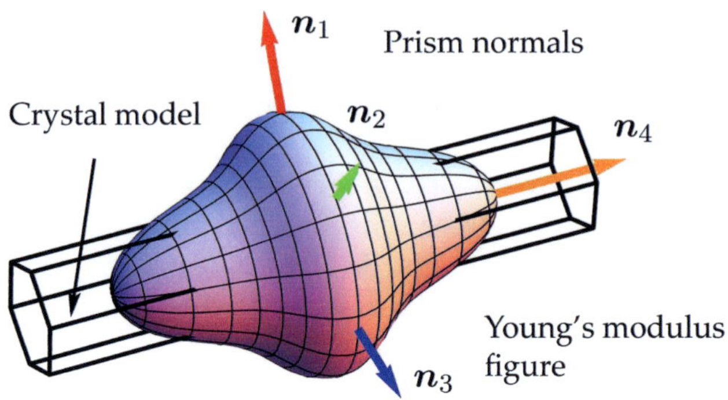

Figure 3.1: Hexagonal prism as grain model together with the spherical projection of the grain stiffness tensor $\mathbb{C}^{\mathcal{G}}$ and prism normal vectors n_i, $i = 1 \ldots 4$

thermal expansion tensors, an active transformation with the grain orientations from the structure generator is carried out.

Vogelgesang et al. (2000) determined the following values for the components of the elastic stiffness tensor by Brillouin scattering at room temperature: $C^{\mathcal{G}}_{11} = 433\,\text{GPa}$, $C^{\mathcal{G}}_{33} = 574\,\text{GPa}$, $C^{\mathcal{G}}_{12} = 195\,\text{GPa}$, $C^{\mathcal{G}}_{13} = 127\,\text{GPa}$, $C^{\mathcal{G}}_{66} = 108\,\text{GPa}$, $C^{\mathcal{G}}_{44} = 119\,\text{GPa}$. The relationship $C^{\mathcal{G}}_{12} = C^{\mathcal{G}}_{11} - 2C^{\mathcal{G}}_{44}$ holds, because only five of the six components are independent in the hyperelastic case. All other stiffness components are zero in this configuration.

Figure 3.1 shows the direction dependent Young's modulus of the stiffness tensor at room temperature together with a lattice model of a β-Si$_3$N$_4$ grain. It is obvious that the grains exhibit a significant degree of anisotropy.

Thermal Strain. The tensorial representation of the thermal expansion coefficients with transverse-isotropic symmetry is given by

$$\boldsymbol{\alpha}^{\mathcal{G}}(T) = \alpha^{\mathcal{G}}_a(T)(\boldsymbol{I} - \boldsymbol{B}_c) + \alpha^{\mathcal{G}}_c(T)\boldsymbol{B}_c, \tag{3.4}$$

where $\boldsymbol{B}_c =: \boldsymbol{B}_3$ defines the axial direction c of the grains and a denotes the basal directions. Henderson and Taylor (1975) determined the thermal expansion of the β-grains by X-ray diffraction methods between 20 and 1020°C. The following quadratic ansatz was used to adjust the experimental results

$$y(T) = y_0\left(1 + k_1 T + k_2 T^2\right), \tag{3.5}$$

where $y(T)$ is the temperature dependent lattice distance. The linear thermal expansion coefficients have been determined from the lattice distance by

$$\alpha(T) = \frac{1}{y(T)} \frac{\mathrm{d}y(T)}{\mathrm{d}T} = \frac{k_1 + 2k_2 T}{1 + k_1 T + k_2 T^2}. \tag{3.6}$$

In the following, the temperature dependence is approximated by a secant interpolation between the values at 20 and 1020°C:

$$\alpha_{a/c}^{\mathcal{G}}(T) = \frac{1}{y_{a/c}(T)} \frac{\mathrm{d}y_{a/c}(T)}{\mathrm{d}T} \approx \alpha_{a/c}^{\mathcal{G}}(T_0) + k_{a/c}^{\mathcal{G}} \Delta T, \tag{3.7}$$

The relation $k_{a/c}^{\mathcal{G}} = (\alpha_{a/c}^{\mathcal{G}}(T_1) - \alpha_{a/c}^{\mathcal{G}}(T_0))/(T_1 - T_0)$ delivers a secant approximation of the temperature dependence. The thermal strain $\varepsilon_{\mathrm{th}}$ can be determined from the thermal expansion coefficient by integration

$$\varepsilon_{\mathrm{th}}(T) = \int_{T_0}^{T} \boldsymbol{\alpha}(\tilde{T}) \, \mathrm{d}\tilde{T}. \tag{3.8}$$

Henderson and Taylor (1975) have delivered the following material parameters: $\alpha_c^{\mathcal{G}}(T_0) = 1.955 \cdot 10^{-6}\mathrm{K}^{-1}$, $\alpha_a^{\mathcal{G}}(T_0) = 0.795 \cdot 10^{-6}\mathrm{K}^{-1}$, $k_c^{\mathcal{G}} = 3.504 \cdot 10^{-9}\mathrm{K}^{-2}$, $k_a^{\mathcal{G}} = 4.846 \cdot 10^{-9}\mathrm{K}^{-2}$. Figure 3.2 a) shows the thermal expansion coefficients for axial and basal direction and b) the relative error e between the definition and the interpolation, see, Eqs. (3.7). This shows that the non-linearity of these quantities is negligible.

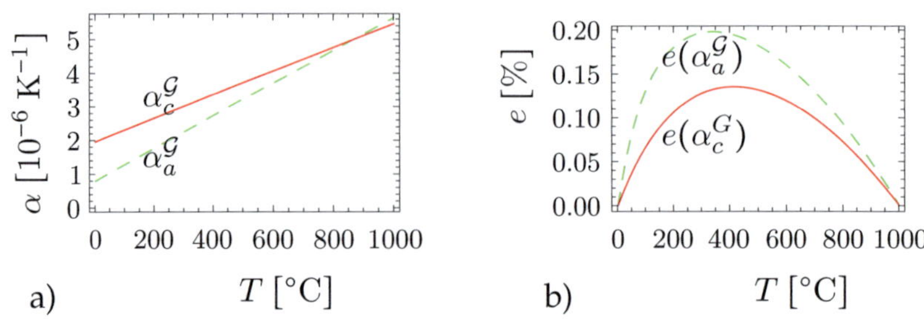

Figure 3.2: a) Thermal expansion of β-crystals, b) relative error between the secant-approximation and the experimental values

3.3.2 Thermoelastic Model for the Glassy Matrix

Stiffness Tensor. The glassy matrix phase ($\mathcal{M}$) is assumed to have an isotropic material behavior. The temperature dependent stiffness tensor then reads

$$\mathbb{C}^{\mathcal{M}}(T) = 3K^{\mathcal{M}}(T)\mathbb{P}_1 + 2G^{\mathcal{M}}(T)\mathbb{P}_2, \tag{3.9}$$

with the bulk modulus $K^{\mathcal{M}}$ and shear modulus $G^{\mathcal{M}}$

$$K^{\mathcal{M}} = \frac{E^{\mathcal{M}}}{3(1 - 2\nu^{\mathcal{M}})}, \quad G^{\mathcal{M}} = \frac{E^{\mathcal{M}}}{2(1 + \nu^{\mathcal{M}})}. \tag{3.10}$$

The volumetric and deviatoric isotropic projectors are $\mathbb{P}_1 = \frac{1}{3}\boldsymbol{I} \otimes \boldsymbol{I}$ and $\mathbb{P}_2 = \mathbb{I}^{\mathsf{S}} - \mathbb{P}_1$, respectively. The variation of Young's modulus at elevated temperatures for several oxynitride glasses can be found, e.g., in Hampshire et al. (1994). For this work, a glass with 17 equiv.% of nitrogen has been chosen in order to represent the glassy phase in the bulk material under consideration. This assumption is made due to the problems implied by the experimental assessment of the material properties in the glassy phase. At room temperature ($T_0 = 20°C$), Young's modulus is

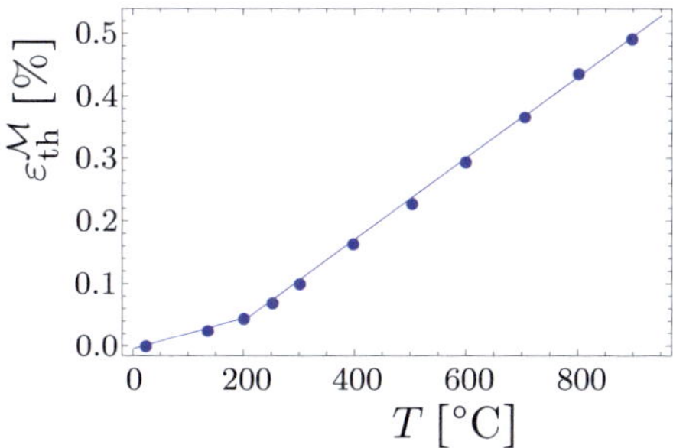

Figure 3.3: Thermal expansion of Y-Sialon glass (with 17% nitrogen and standard cation composition). The dots correspond to the experimental data by Hampshire et al. (1994). The solid line is a bilinear approximation. The data is used as model for glassy phase on the microscale.

$E^{\mathcal{M}}(T_0) = 133\,\mathrm{GPa}$. At $T_1 \approx 800°\mathrm{C}$, Young's modulus has decreased by approximately 8.6% (Hampshire et al., 1994). The variation is linear in this temperature range, such that Young's modulus can also be expressed by a secant approximation

$$E^{\mathcal{M}}(T) \;=\; E^{\mathcal{M}}(T_0) + k^{\mathcal{M}}\Delta T, \tag{3.11}$$

$$k^{\mathcal{M}} \;=\; \frac{E^{\mathcal{M}}(T_1) - E^{\mathcal{M}}(T_0)}{T_1 - T_0}. \tag{3.12}$$

Due to the lack of experimental data, Poisson's ratio is assumed to have a constant value of $\nu^{\mathcal{M}} = 0.29$, which is motivated by (Hampshire et al., 1994).

Then the stiffness tensor has the final representation

$$\mathbb{C}^{\mathcal{M}}(T) = E^{\mathcal{M}}(T)\left(\frac{\mathbb{P}_1}{1 - 2\nu^{\mathcal{M}}} + \frac{\mathbb{P}_2}{1 + \nu^{\mathcal{M}}}\right). \tag{3.13}$$

Thermal strain. The thermal expansion for isotropic materials is denoted by $\boldsymbol{\alpha}(T) = \alpha(T)\boldsymbol{I}$ or $\boldsymbol{\varepsilon}_{\text{th}}(T) = \varepsilon_{\text{th}}(T)\boldsymbol{I}$ with $\varepsilon_{\text{th}}(T) = \int_{T_0}^{T} \alpha(\tilde{T})\, \mathrm{d}\tilde{T}$. Values for the expansion of oxynitride glasses have been documented by Hampshire et al. (1994). The thermal expansion coefficient is approximately constant at a value of $\alpha_0^{\mathcal{M}} = 2.5 \cdot 10^{-6}\text{K}^{-1}$ between $T_0 = 20°\text{C}$ and $T_1 = 210°\text{C}$. Above T_1 it has approximately the value $\alpha_1^{\mathcal{M}} = 6.5 \cdot 10^{-6}\text{K}^{-1}$ (see Figure 3.3). Hence, the thermal strain can be described by the piecewise linear function

$$\varepsilon_{\text{th}}^{\mathcal{M}}(T) = \begin{cases} \alpha_0^{\mathcal{M}}(T - T_0) & \text{if} \quad T_0 \leqslant T < T_1 \\ \alpha_0^{\mathcal{M}}(T_1 - T_0) + \alpha_1^{\mathcal{M}}(T - T_1) & \text{if} \quad T_1 \leqslant T \end{cases} . \tag{3.14}$$

Such a piecewise linear approximation is in some contradiction with respect to the assumption of a continuous differentiable thermoelastic strain energy function. A continuous approximation could easily be generated, which would introduce more parameters. But the non-differentiability is not of any importance in this context.

3.4   Finite Element Model

3.4.1   Geometrical Model

The finite element calculations have been carried out on periodic ensembles with 64 and 238 grains. These ensembles were generated by the self-written microstructure generator, which was described in Chapter 2.

Structures with mean grain length of around 12.2 and 7.3 µm and mean aspect ratio of 3.4 and 3.0 for the 64 and 238 grain ensemble, respectively, could thereby be obtained.

Figure 3.4: Example for a periodic microstructure in high resolution; for a better insight into the structure, the grain volume fraction has been reduced.

The visual comparison of silicon nitride and the generated microstructure shows a feasible agreement (Figure 2.11). A certain grain overlapping can be seen in both the micrograph and the artificial structure.

Figure 3.4 shows an example for the three-dimensional microstructures after discretization in high resolution with the commercial meshing tool ScanIP from Simpleware (Young et al., 2008). The grain volume fraction has been reduced in order to give a better insight into the three-dimensional microstructure. It is characterized by elongated grains with rounded grain boundaries and by an isotropic orientation distribution with full periodicity.

The number of degrees of freedom (DoF) has been varied in the range of 5×10^5 and 1×10^6. The thermoelastic constitutive equations have been implemented in Abaqus/Standard into the user material routine UMAT (Hibbitt et al., 2001).

3.4.2 Projected Periodic Boundary Conditions

The periodicity of the deformation obtained using periodic displacement fluctuation boundary conditions allows for an efficient calculation of the effective thermoelastic properties of the microheterogeneous material (Suquet, 1982). They avoid stress concentrations near the boundary of the unit cell, which are due to over-constraints of the displacement field near the boundary of the considered volume element in the case of homogeneous boundary conditions. The displacement field on the boundary $u^\pm$ can be decomposed additively into the periodic fluctuations $w^\pm = u^\pm - \bar{\varepsilon} x^\pm$ and the homogeneous displacement due to the prescribed mean strain $\bar{\varepsilon}$ according to $w^+ = w^-$.

Usually, periodic boundary conditions (PBCs) are implemented in finite element formulations either in combination with regular meshes or with meshes, which are preserving the periodicity of the geometry in the discretization. The first approach implies that the geometric information is approximated in a quite coarse manner or a prohibitively fine mesh has to be used. The second approach see, e.g., Flaquer et al. (2007) or Fritzen et al. (2009); Fritzen and Böhlke (2010) is much more efficient. Here, a tetrahedral mesh is modified on the boundaries to a conformal mesh. It is geometrically exact and allows for the description of complicated microstructures with a comparable small number of degrees of freedom. The disadvantage is the complicated implementation due to the necessity of an extension of an existing mesh generator.

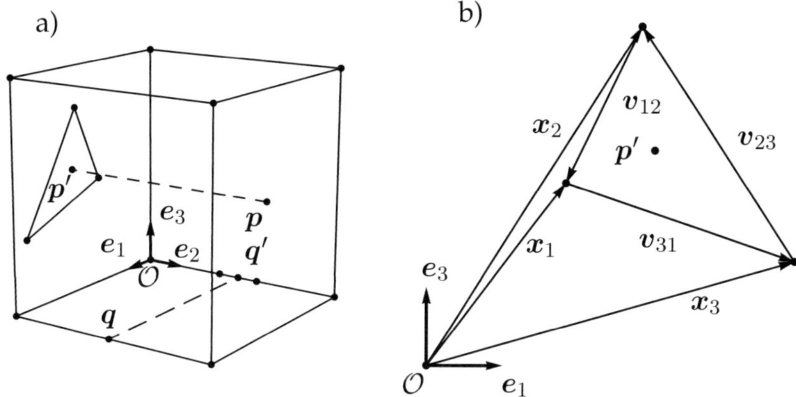

Figure 3.5: a) Cube with triangle-to-node and segment-to-node relations, b) triangle with vectors

A further option is the approximation of periodicity of the discretization by geometric considerations. This approach will be used in the following and represents one of the important technical results of the paper, because it allows for an efficient and accurate estimation of the thermoelastic behavior with limited sizes of the considered unit cells.

So, the node-wise correspondence can be weakened in a triangle-to-node relationship on the faces and to a segment-to-node relationship on the edges, see Figure 3.5 a). In order to find corresponding triangle-node sets, it has to be distinguished, whether a node projection is inside or outside of a triangle. This can be accomplished by a consideration of the norm $\mathcal{N}$ of the triangle, which is given by

$$\mathcal{N} = \max \left\{ \frac{\boldsymbol{n}_{ij} \cdot (\boldsymbol{x}_\mathrm{m} - \boldsymbol{p}')}{\boldsymbol{n}_{ij} \cdot \boldsymbol{n}_{ij}} \right\}, \tag{3.15}$$

with $ij = 12, 23, 31$, where $\boldsymbol{n}_{ij}$ are the midpoint based normals on the edges. They can be calculated using the midpoint $\boldsymbol{x}_\mathrm{m}$ and the edge vectors $\boldsymbol{v}_{ij}$ of the triangle as depicted in Figure 3.5 b):

$$\boldsymbol{n}_{ij} \;=\; \boldsymbol{x}_{\mathrm{m}} - \boldsymbol{x}_i - \frac{(\boldsymbol{x}_{\mathrm{m}} - \boldsymbol{x}_i)\cdot \boldsymbol{v}_{ij}}{\boldsymbol{v}_{ij}\cdot \boldsymbol{v}_{ij}}\boldsymbol{v}_{ij}, \tag{3.16}$$

$$\boldsymbol{x}_{\mathrm{m}} \;=\; \frac{1}{3}(\boldsymbol{x}_1 + \boldsymbol{x}_2 + \boldsymbol{x}_3) \tag{3.17}$$

and $\boldsymbol{v}_{ij} = \boldsymbol{x}_i - \boldsymbol{x}_j$. If $0 \leqslant \mathcal{N} \leqslant 1$, the node projection $\boldsymbol{p}'$ is in the triangle or on the edges. This point can be expressed in terms of the triangle vectors, e.g., as

$$\boldsymbol{p}' = \boldsymbol{x}_1 + s\boldsymbol{v}_{12} - t\boldsymbol{v}_{31}, \tag{3.18}$$

such that the equation for the projected PBCs on the faces is

$$\boldsymbol{p}' - \bar{\varepsilon}\boldsymbol{x}^{+} = (1 - s + t)\,\boldsymbol{x}_1 + s\,\boldsymbol{x}_2 - t\,\boldsymbol{x}_3 - \bar{\varepsilon}\boldsymbol{x}^{-}. \tag{3.19}$$

The edges have to be considered in a similar manner. The corresponding equation is finally denoted as

$$\boldsymbol{q}' - \bar{\varepsilon}\boldsymbol{x}^{+} = (1 - r)\,\boldsymbol{x}_1 + r\,\boldsymbol{x}_2 - \bar{\varepsilon}\boldsymbol{x}^{-}. \tag{3.20}$$

The mean deformation $\bar{\varepsilon}$ is always prescribed on the corners of the corresponding faces or edges of the unit cell.

3.4.3 Determination of Effective Thermoelastic Material Properties

For the estimation of the macroscopic elastic stiffness tensor at room temperature and at elevated temperatures, six orthogonal deformation modes have to be applied to the structures. The deformations are shown in Figure 3.6.

The periodicity of the mechanical fields can be seen in all cases. A comparison of the opposing edges of the unit cells shows the

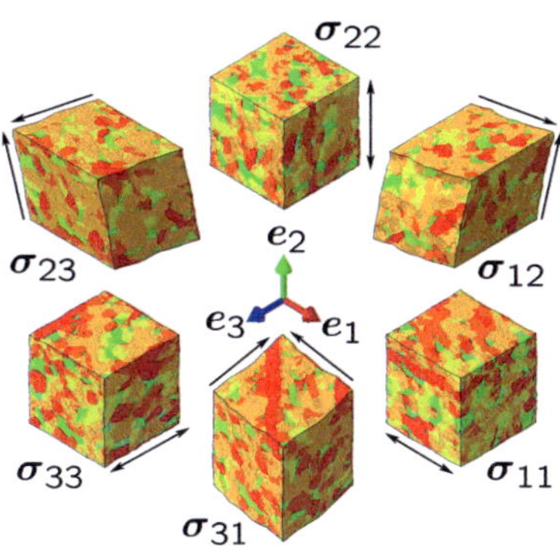

Figure 3.6: Six deformation modes on a periodic unit cell with 232 grains. The arrows indicate the deformation directions. The stress distributions are inhomogeneous and periodic.

effect of the periodic boundary conditions, visualized by an over-pronounciation of the deformation field.

The effective elastic properties can be obtained, if the α-th component of the effective stress tensor $\langle \sigma^\alpha \rangle$ due to the strain mode $\bar{\varepsilon}^\beta = \varepsilon_0 \boldsymbol{B}_\beta$ is divided by the components of the effective strain tensor, so that $\bar{C}_{\alpha\beta} = \langle \sigma^\alpha \rangle \cdot \boldsymbol{B}_\beta / \varepsilon_0$. The effective stress tensor is given by the volume averages of the corresponding local fields.

The uniqueness of the temperature dependence parameter of the stiffness $k^{\mathcal{G}}$ can be seen, if the volume averages of stress with the thermoelastic strain $\varepsilon^*_{\text{th}} := \varepsilon - \varepsilon_{\text{th}}$,

$$\langle \boldsymbol{\sigma} \rangle = \frac{1}{V} \int_V \mathbb{C}(T)[\varepsilon^*_{\text{th}}] \, \mathrm{d}V \tag{3.21}$$

is considered in detail. It contains contributions from the glassy phase ($\mathcal{M}$, Eq. 3.9) and from each grain ($\mathcal{G}_i$, Eq. 3.2).

$$V \langle \boldsymbol{\sigma} \rangle = \int_{V^{\mathcal{M}}} \mathbb{C}^{\mathcal{M}}(T)[\varepsilon^*_{\text{th}}] \mathrm{d}V^{\mathcal{M}} + (1 + k^{\mathcal{G}} \Delta T) \sum_{i=1}^{n^{\mathcal{G}}} \left(\int_{V_i^{\mathcal{G}}} \mathbb{C}^{\mathcal{G}}_{0\,i}[\varepsilon^*_{\text{th}}] \, \mathrm{d}V_i^{\mathcal{G}} \right) \tag{3.22}$$

This equation evidently allows for a unique condensation of the parameter $k^{\mathcal{G}}$.

The effective thermal expansion tensor $\bar{\alpha}$ is obtained by the derivative of the effective strain tensor with respect to the temperature, i.e., $\bar{\alpha}(T) = \partial_T \langle \varepsilon_{\mathrm{th}} \rangle$. In an incremental setting with sufficiently small ΔT, the expansion coefficient is given by $\bar{\alpha}(T) \approx \Delta \langle \varepsilon_{\mathrm{th}} \rangle / \Delta T$.

3.5   Numerical Results

3.5.1   Effective Stiffness and Size of the Unit Cell

The macroscopic thermal expansion coefficient and Young's modulus have been determined for the commercial silicon nitride grade SL 200 BG within the ESIS project (Lube and Dusza, 2007). Young's modulus decreases from 302.9 GPa at room temperature to 294.3 GPa at 800°C.

The homogenization of the elastic stiffness has been carried out at room temperature (20°C) and the elevated temperature 800°C. Table 3.1 arranges the isotropic elastic constants, which are calculated from the effective stiffness tensor $\bar{\mathbb{C}}$. The symbols correspond to Figure 3.7. The consideration of the eigenvalues of an isotropic stiffness tensor with $3\bar{K} = \bar{\mathbb{C}} \cdot \mathbb{P}_1 / \|\mathbb{P}_1\|^2$ and $2\bar{G} = \bar{\mathbb{C}} \cdot \mathbb{P}_2 / \|\mathbb{P}_2\|^2$ allows for the calculation of the effective engineering constants Young's modulus $\bar{E}$ and $\bar{\nu}$ by $\bar{E} = 9\bar{K}\bar{G}/(3\bar{K} + \bar{G})$ and $\bar{\nu} = (3\bar{K} - 2\bar{G})/(2(3\bar{K} + \bar{G}))$.

Figure 3.7 shows the values of Table 3.1 together with experimental data from Lube and Dusza (2007). The linear isotropic decrease of the β-Si_3N_4 stiffness components is determined to 4.42% between room temperature and 1000°C, which corresponds to a factor $k^{\mathcal{G}} = -4.29 \cdot 10^{-5} \mathrm{K}^{-1}$.

Table 3.1: Effective isotropic elastic material properties, symbols correspond with Figure 3.7; DoF denotes the degrees of freedom of the used finite element discretization; the effective Poisson's ratio is $\bar{\nu} = 0.283$ for all considered cases.

Sym.	Grains	DoF $[10^6]$	$T\,[^\circ\mathrm{C}]$	$\bar{K}\,[\mathrm{GPa}]$	$\bar{G}\,[\mathrm{GPa}]$	$\bar{E}\,[\mathrm{GPa}]$
▪	64	0.49	20	234.9	119.5	306.6
●	64	1.03	20	233.9	119.0	305.3
▪	64	0.49	800	227.5	115.6	296.6
●	64	1.03	800	226.9	115.4	295.9
◆	238	0.65	20	236.9	120.3	308.7
◆	238	0.65	800	229.6	116.5	299.0

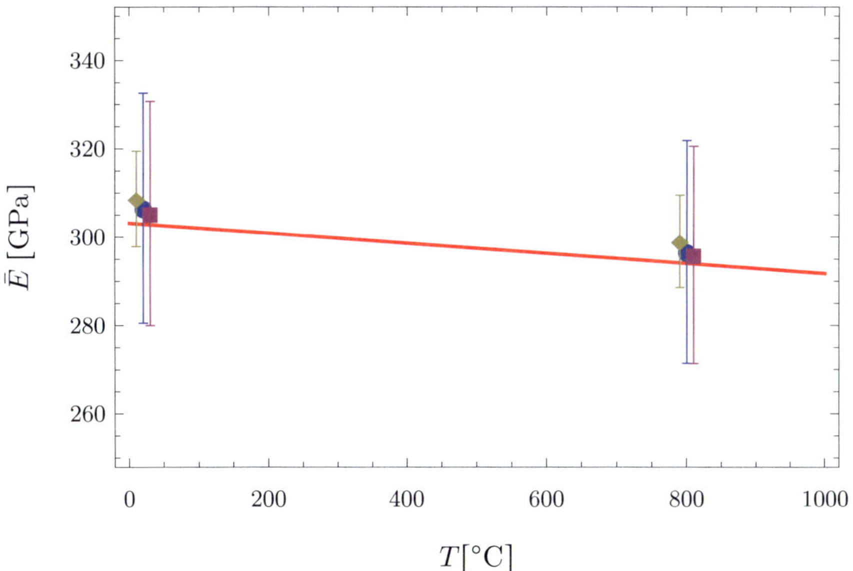

Figure 3.7: Young's modulus from experiment of Lube and Dusza (2007) (−) and unit cells with 64 grains (▪, ●) and 238 (◆) grains

The isotropic Young's modulus can be compared to the directional Young's modulus $E(\vartheta, \varphi)$ see, e.g., Böhlke and Brüggemann (2001), in order to examine the representativity of the investigated unit cells. This can be done by consideration of the normalized difference between the two quantities: $\Delta E(\vartheta, \varphi) = (E(\vartheta, \varphi) - \bar{E})/\bar{E}$. The error bars in Figure 3.7 indicate the minimum and maximum values of the directional values. The maximum amplitude of $|\Delta E(\vartheta, \varphi)|$ is 8.4% for the 64 grain ensemble, and 3.5% for the 238 grain ensemble. A slight effect due to mesh density can be seen between the two values for the smaller ensemble: The value of $\bar{E}$ for the coarser mesh (∎) is 0.41% higher than the one for the finer mesh (•). The small overestimation of the data by the simulated values can be attributed to uncertainties of the volume fraction of the materials and of the assumed properties of glassy phase material.

3.5.2 Effective Thermal Expansion

The thermal expansion of the bulk material shows a significant non-linear behavior (Figure 3.8). This data, as well as the recent measurements, serve as comparison values for the validation of the used homogenization.

In order to trace the non-linear temperature dependence of the thermal strains between room and elevated temperature, the isothermal cooling process was discretized into ten time steps. The results of the simulations have been compared to experimental data for different grades of silicon nitride from present experiments (Section 3.2) and from literature (Lube and Dusza, 2007).

The experimental data and the simulated curves of the isotropic part of the thermal expansion coefficient $\bar{\alpha}(T)$ with $\bar{\alpha}(T) = \nicefrac{1}{3} \operatorname{tr}(\bar{\boldsymbol{\alpha}}(T))$ are shown in Figure 3.8: The subfigures a), c) and e) compose simulated values of the effective thermal expansion with experimental values for AlY (SL 200 BG), MgLu and MgY as sintering additives. The subfigures b), d) and f) show thermal expansion coefficients corresponding to the thermal expansion curves in the left column.

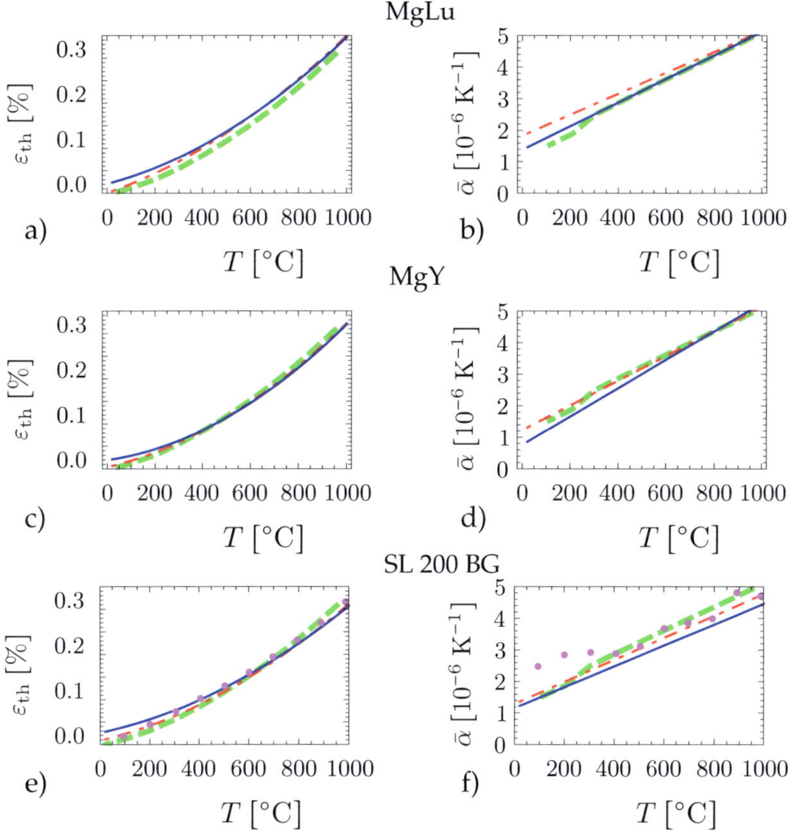

Figure 3.8: Simulated (- - -) and experimental values for the effective thermal expansion (a, c, e) and the effective thermal expansion coefficients (b, d, e) of Lube and Dusza (2007) (•) and from present experiments measured during heating (- · -) and cooling (—).

The non-linear temperature dependence of the thermal expansion in Figure 3.8 a), c) and e) in both the experimental and the simulated curves corresponds directly to the non-linearity of the microscopic results of Henderson and Taylor (1975), which have been used for the derivation of the thermal expansion coefficients in Eqs. (3.5) and (3.6). The influence of the non-linearity of the thermal expansion of the assumed glassy phase (Eq. (3.14)) on the effective behavior is almost negligible, due to its low volume fraction of 12%. It is best visible as small kink in the thermal expansion coefficient in Figure 3.8 b), d) and f).

The difference between simulated values and experimental results is reasonably small. Similarly to the Young's modulus, the difference between the homogenized directional representation of the thermal expansion coefficient $\Delta\alpha(T,\vartheta,\varphi) = (\alpha(T,\vartheta,\varphi) - \bar{\alpha}(T))/\bar{\alpha}(T)$ can be used to evaluate the homogeneity of the obtained effective thermal expansion coefficient.

At room temperature, the anisotropy of the effective thermal expansion coefficient is more pronounced than at higher temperatures, which is a result of the increasing anisotropy of the thermal expansion coefficients of the β-grains at lower temperatures, compare Figure 3.2 a). At room temperature, the difference is 13.5% for the cell with 64 grains, and 9.5% for the cell with 238 grains. At 900°C, the difference is approximately 8.1% for the cell with 64 grains and 5.5% for the cell with 238 grains. Thus, the same trend as for the effective Young's modulus is observed. Interesting is the fact that the anisotropy of the thermal expansion coefficients at room temperature is higher than the ansiotropy of the elasticity tensor.

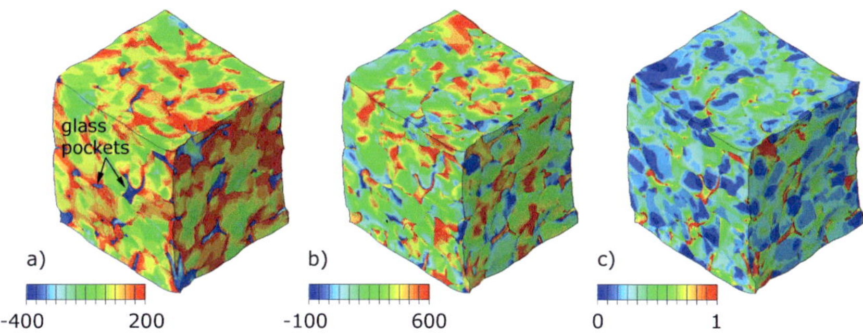

Figure 3.9: Periodic ensemble with 238 grains: a) hydrostatic pressure, b) maximum principal stress, c) elastic strain energy; all values in [MPa]

3.5.3 Residual Stress Distribution within the Unit Cell

For a better understanding of the influence of the thermal strains on the material behavior, it is useful to consider the local field solutions of the finite element simulations. The following figures show the hydrostatic pressure, the maximum principal stress and the elastic strain energy density after the cooling-down process from the glass transition temperature. Above this temperature, a stress relaxation is assumed. Additionally, three different measures for stress triaxiality are shown for comparison.

Figure 3.9 a) shows the hydrostatic pressure field. The glass pockets are exposed to significant hydrostatic tension, whereas the grains are mainly under slight compression. The subfigures 3.9 b) and c) show two quantities, which are expected to act as driving forces for the crack propagation in brittle materials. The maximum principal stress in Figure 3.9 b) can be used as fracture criterion. Here, the matrix pockets and the grain boundaries are under significant tensile stresses. The elastic strain energy, Figure 3.9 c), indicates, which locations are most likely to initiate micro cracks. These regions can be found within the glassy pockets and near the grain boundaries, which are exposed to both pronounced tensile stresses and high elastic energy densities. A comparison between the computed stress fields and the results from Peterson and Tien (1995) for silicon nitride with 88% β-crystals using the Eshelby inclusion method show a good agreement. The periodicity of the deformations and with it the periodicity of the depicted static quantities can be observed in all cases.

The triaxiality measures are especially of interest, if silicon nitride is exposed to elevated temperatures above the glass transition temperature, where plastic flow plays a prominent role for the description of high temperature effects like creep or stress relaxation. The first considered triaxiality measure is the classical triaxiality $\mathcal{T} = -p/\sigma_{\mathrm{eq}}$, where p is the hydrostatic pressure and σ_{eq} is the von Mises stress. For uniaxial tensile stress is $\mathcal{T} = 1/3$. A higher positive value means higher amplitude of hydrostatic tension without driving force for stress reduction due to plastification after the von Mises yield criterion. A clear drawback is the incapability of the distinction between generally deviatoric stress fields and pure shear: In both cases is $\mathcal{T} = 0$. A second option for the description of multi-axial stress fields is the Lode parameter

$$\mu = \frac{2\sigma_2 - \sigma_1 - \sigma_3}{\sigma_1 - \sigma_3}, \tag{3.23}$$

with the maximum, intermediate and the minimum principal stresses σ_1, σ_2 and σ_3, respectively (Lode, 1926). The intermediate stress plays a prominent role here. The Lode parameter varies between -1 and 1. Values around ± 1 indicate a general compressive/tensile state, respectively. Shear stresses are characterized by vanishing values. The third quantity is the determinant or third invariant of the normalized deviatoric stress with

$$\widetilde{III}^* = 3\sqrt{6}\det\left(\sigma'/\|\sigma'\|\right), \tag{3.24}$$

which is closely related to the Lode angle. Like the Lode param-

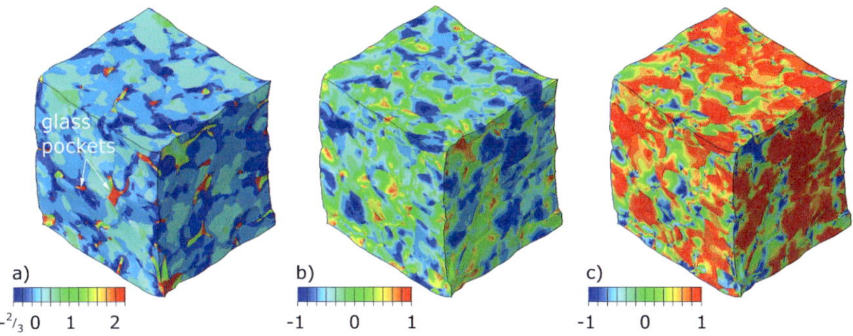

Figure 3.10: Periodic ensemble with 238 grains: a) the classical tri-axiality $\mathcal{T}$, b) the Lode parameter μ and c) the third invariant of the normalized deviatoric stress $\widetilde{III}^{*}$

eter, $\widetilde{III}^{*}$ has fixed bounds. It varies between ± 1, where 1 means uniaxial elongation, 0 is plane strain compression, and -1 is simple compression see, e.g., Böhlke and Bertram (2001) or Wippler and Böhlke (2010).

The main difference is the switch of signs. The equation $\mu = -\widetilde{III}^{*}$ has seven solutions. Three are simple and quite obvious. If the intermediate stress σ_2 is searched for, the equation has solutions in $\sigma_2 = \sigma_1$, $\sigma_2 = \sigma_3$ and $\sigma_2 = \frac{1}{2}(\sigma_1 + \sigma_3)$. These solutions are linked to the parameter values ± 1 and 0. The other four solutions have a complicated structure and result from a fourth-order polynomial.

Figure 3.10 a) shows clearly that the glass pockets are undergoing a high stress triaxiality, due to the hydrostatic tension (see Figure 3.9). The grains are under compression, such that they see a slight negative stress triaxiality. The subfigures 3.10 b) and c) show the Lode parameter and the third invariant of the normalized deviatoric stress. As mentioned above, they show a similar picture with exchanged signs. The grain boundaries are characterized by a jump from positive to negative sign of the two quantities, indicating shear stress gradients in these locations.

3.6   Summary and Conclusions

The magnitude of the stress fields, which have been obtained by the finite element method is in agreement with the predictions from the Eshelby method of Peterson and Tien (1995).

The advantage of the consideration here is that full-field results are obtained. The local fields of thermal strains, stresses and triaxiality measures give deeper insight into the self-loading of silicon nitride due to cooling down from sintering temperature.

The numerically obtained temperature dependent effective material parameters show a feasible representation of the experimentally observed effective material parameters.

The material behavior on the microlevel had to be adjusted due to missing information on the temperature dependence of the stiffness tensor for β-Si_3N_4. So, a linear dependency on a single parameter was chosen, which has been determined by solving an inverse problem. The uniqueness of the inverse procedure was shown by a consideration of the volume averages, which have been used for the calculation of the effective stiffness tensor.

The thermal expansion for the glassy phase has been modeled piecewise linear according to the experimental observations. The non-linearity of the grains' thermal expansion was modeled by a quadratic ansatz. With these ingredients, it was possible to calculate the effective thermal expansion of bulk silicon nitride from microstructural information.

Hence, the thermoelastic modeling based on temperature dependent data was successful, hereby, giving a complete thermoelastic model of silicon nitride. Due to the relatively low number of grains in the ensembles, some deviations from isotropy in mainly single-digit percentage have been observed. The trend to convergence for an increasing number of grains is clearly visible.

Based on the numerical results on the stress fields it can be concluded that the most endangered regions are the glass pockets and the grain boundaries, because here high stress and strain energy concentrations are arising, which can lead to cracking on the microscale during the cooling phase.

This implies that the model can be applied for further considerations such as fracture simulations under thermal influences. So, the thermal strains and stresses definitively have to be considered for the fracture behavior of silicon nitride, which will be addressed in the next chapter.

The implementation of the periodic boundary conditions in an approximative setting using a projection algorithm allows for an efficient mapping microstructure by relatively low numerical costs. They avoid the undesirable effects of uniform kinematic and static boundary conditions. Therefore, they appear as a versatile approach for the determination of effective material behaviors in the context of complex microstructures. To the best knowledge of the authors this is the first implementation of periodic boundary conditions in such a format.

Chapter 4

Fracture Behavior

4.1 Introduction

In this chapter, the fracture properties of silicon nitride for the onset of cracking will be examined. The aim of the work is the formulation of an effective material behavior, which is based on micromechanical considerations. The approach is widely used in engineering mechanics, when an efficient description of a micro-heterogenous material is desired.

In particular, this means that the geometric and thermoelastic material information on the microscopic scale has to be incorporated into one detailed model (Maugin, 1992). So, the results from the Chapters 2 and 3 are extended with assumptions on the fracture behavior for all relevant constituent components.

As it has been already mentioned, the sinter material consists mainly of β-Si_3N_4 grains, which give high strength and stiffness to the bulk material. The glassy phase appears in two different morphologies: First, it forms glass pockets at grain-free material volume and second, it develops intergranular films between the grains. Both are crucial for the fracture behavior. The first form is relevant for the residual stresses in the material, which are important for the formation of bridging grains, as it is pointed out in the work of Peterson and Tien (1995) (see as well, Sections 3.1 and 3.5.3). Kruzic et al. (2008) describes in detail how the properties of the second manifestation of the glassy phase are crucial for the efficient exploitation of the bridging grains and with it for the effective fracture properties of silicon nitride: Relatively low strength of the intergranular films lead to toughening due to crack path deviation, i.e., cracks have to propagate around the strong grains. So, the crack propagation consumes more energy; a tougher material with relatively low fracture strength is the consequence. On the other hand, high interface strength leads to an increased fracture strength at comparable low fracture toughness. Hence, a good compromise has to be found, in order to obtain optimum overall fracture properties.

A comprehensive constitutive model has to be created for each fracture mechanism on the microscale, which incorporates all features of importance, i.e., fracture of the brittle glass pockets and the grains under maximum normal stresses and the delamination of the grain boundaries under tensile and shear tractions.

Given the microscopic models, the effective fracture behavior can be determined. Although such effective behavior in general can be determined by rigid methods, this approach is difficult – not to say – practically impossible for a complex material behavior, because the effective material would need an infinite number of internal variables for the description of the material state evolution, as it has been pointed out in Maugin (1992).

The calculation of effective fracture properties for a brittle material with growing microcracks is a topic of ongoing research. A thermodynamical rigid method for the estimation of effective properties in a solid material with growing cracks has been proposed by

Costanzo et al. (1996). Here, fourth-order localization tensors are used for the determination of the effective solution. However, the method is using several assumptions, which are not valid for the considered case of silicon nitride. Nevertheless, the method yields the inspiration for a consideration of the stress power. This concept will be used to show the energetic consistence of the proposed volume averaging technique. More relevant for a practical implementation for the determination of the fracture behavior in brittle solids with complex microstructure due to adhesive and cohesive failure is the work of Verhoosel et al. (2010). Here, a FE^2-approach was used to model a composite material in a two-dimensional framework. Although the approach appears promising due to the fact that the FE^2-method is only applied in the domains, which are exposed to fracture, the numerical costs are definitely to high for a three-dimensional application.

So, another way had to be found, which is conceptually simpler and numerically feasible for the considered case of silicon nitride. Thus, a phenomenological model has been identified, which incorporates the main features of the observable macroscopic fracture behavior and which allows for the subsequent determination of a limited number of fracture properties from the effective results of unit cell simulations.

4.2 Theoretical Aspects of Local and Effective Material Behavior

As already mentioned, is the description of the effective material behavior by an averaging method over all local material behaviors on the microscale the primary aim of this work. The chosen method in this section is based on the unit cell approach, which has been introduced in Section 3.5. In the case of thermoelasticity,

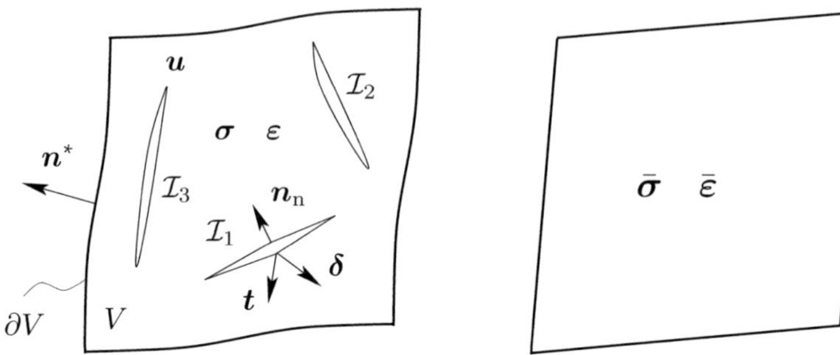

Figure 4.1: Volume element on the microscopic and macroscopic level with corresponding quantities

the balance of microscopic and macroscopic stress power is a priori fulfilled, as it is well known, since the fundamental work of Hill (1963). For a material with growing cracks and load transfer over the crack flanks, in addition to the purely linear elastic stress power, contributions from those cracks have to be incorporated.

In Figure 4.1 a portion of the considered material can be seen on both the micro- and the macroscale. The left side shows the microheterogeneous material with local stress and strain fields σ and ε. Additionally, cracks are visible, which are designated as interfaces $\mathcal{I}$. On these interfaces traction vectors $t = \sigma n_{\mathrm{n}}$ (Lemma of Cauchy) and separations, defined as displacement jumps $d = [\![u^+ - u^-]\!] \; \forall \; x \in \mathcal{I}$ on the singular surfaces $\mathcal{I} \in V$ are acting.

The right side of Figure 4.1 shows the same portion of the material after the averaging procedure. Here, only the effective quantities $\bar{\sigma}$ and $\bar{\varepsilon}$ are visible. For sake of thermodynamic consistency of the model, it is important to consider the balance of the effective stress power $\bar{\mathscr{P}}$ and the stress power $\mathscr{P}^\mu$ on the microscale. The Hill-Mandel condition (Hill, 1963) delivers a framework, which allows for this considerations and can be denoted by

$$\bar{\mathscr{P}} = \mathscr{P}^{\mu}.$$ (4.1)

The effective stress power is defined by

$$\bar{\mathscr{P}} = \bar{\boldsymbol{\sigma}} \cdot \dot{\bar{\boldsymbol{\varepsilon}}}$$ (4.2)

and the stress power on the microscopic level is motivated by the work of Costanzo et al. (1996)

$$\mathscr{P}^{\mu} = \langle \boldsymbol{\sigma} \cdot \dot{\boldsymbol{\varepsilon}} \rangle + \frac{1}{V} \sum_{\xi=1}^{n^{\mathcal{I}}} \oint_{\mathcal{I}_{\xi}} \boldsymbol{t} \cdot \dot{\boldsymbol{d}} \, \mathrm{d}A.$$ (4.3)

The effective stress $\bar{\boldsymbol{\sigma}}$ can be identified by the volume average or its boundary integral form due to the Theorem of Gauss

$$\langle \boldsymbol{\sigma} \rangle = \frac{1}{V} \int_{V} \boldsymbol{\sigma} \, \mathrm{d}V = \frac{1}{V} \oint_{\partial V} \mathrm{sym}(\boldsymbol{\sigma} \boldsymbol{n}^{*} \otimes \boldsymbol{x}^{*}) \, \mathrm{d}A$$ (4.4)

with the normal vector $\boldsymbol{n}^{*}$ and the locations $\boldsymbol{x}^{*}$ on the external boundary ∂V, see, Figure 4.1. So, the internal tractions on the interfaces $\mathcal{I}$ do not enter the effective stress, such that $\bar{\boldsymbol{\sigma}} \equiv \langle \boldsymbol{\sigma} \rangle$ holds. The effective strain $\bar{\boldsymbol{\varepsilon}}$ needs a more detailed consideration. Similarly to the effective stress $\bar{\boldsymbol{\sigma}}$, it can be determined by a boundary integral from the local displacement field $\boldsymbol{u}$, see, e.g., Hill (1963). In general, there are contributions from the displacement in the continuum $\boldsymbol{u}^{V} = \langle \boldsymbol{\varepsilon} \rangle \boldsymbol{x}$, from the displacement fluctuations in the continuum $\boldsymbol{w}^{V}$ due to the micro heterogeneity of the material with its restrictions due to the periodic boundary conditions (Suquet, 1982) and from the interface delamination, which is represented by the variable $\boldsymbol{u}^{\mathcal{I}}$. Together,

$$\boldsymbol{u} = \boldsymbol{u}^{V} + \boldsymbol{w}^{V} + \boldsymbol{u}^{\mathcal{I}}$$ (4.5)

is obtained. In this relation, the interesting part is the displacement field due to the interface delamination $\boldsymbol{u}^{\mathcal{I}}$. It is thought as

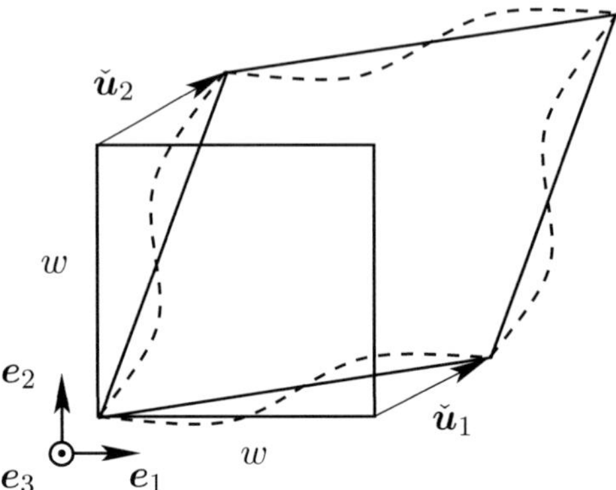

Figure 4.2: Kinematic of a unit cube, examplified on a projection into the e_1-e_2-plane

a generalization of the Heaviside function concept, which has been proposed by Simo et al. (1993) and used by Verhoosel et al. (2010) for a single crack to the considerably more challenging case of a crack pattern, which is not known in detail. The variable $u^{\mathcal{I}}$ has in general an extremely complicated form, which is not accessible by analytical standard techniques. Due to this fact, this issue is not followed further in the context of this work. In case of a unit cube with periodic boundary conditions, the effective strain $\bar{\varepsilon}$ can be determined from the effective displacement gradient $\bar{H}$ by the consideration of three displacement vectors of the unit cube corners $\check{u}_i$ with the coordinates $\check{x}_i$ and $i = 1 \ldots 3$ (Verhoosel et al., 2010). Figure 4.2 shows the kinematic relationships for a projection of a unit cube into the e_1-e_2-plane. The solid square represents the original configuration. The dashed sketch shows a general case for a periodic displacement with stretch and shear contributions and the solid rhomboid is the original square deformed by the effective displacement gradient $\bar{H}$. So, following representation is used:

$$\check{u}_i = \bar{H}\check{x}_i \quad \Leftrightarrow \quad o = \bar{H}\check{x}_i - \check{u}_i, \tag{4.6}$$

where $\boldsymbol{o}$ is the zero vector. An application of the linear map allows for the transformation

$$0 = \left(\bar{\boldsymbol{H}} - \sum_{i=1}^{3} \frac{\check{\boldsymbol{u}}_i \otimes \check{\boldsymbol{x}}_i}{\check{\boldsymbol{x}}_i \cdot \check{\boldsymbol{x}}_i} \right) \check{\boldsymbol{x}}_i, \quad \check{\boldsymbol{x}}_i := w\boldsymbol{e}_i \neq \boldsymbol{o}, \tag{4.7}$$

which delivers the effective deformation gradient as function of the cube size and the corner displacement by

$$\bar{\boldsymbol{H}} = \frac{1}{w} \sum_{i=1}^{3} \check{\boldsymbol{u}}_i \otimes \boldsymbol{e}_i. \tag{4.8}$$

The infinitesimal strain tensor is the symmetric part of the displacement gradient, such that

$$\bar{\boldsymbol{\varepsilon}} = \frac{1}{w} \sum_{i=1}^{3} \mathrm{sym}(\check{\boldsymbol{u}}_i \otimes \boldsymbol{e}_i) \tag{4.9}$$

holds. The displacement on the corners $\check{\boldsymbol{u}}_i$ can by described by a simplified version of the general displacement formula from Eq. (4.5). The displacement fluctuation $\boldsymbol{w}$ vanishes on the corners due to the definition of the periodic boundary conditions (Section 3.4.2). Displacements due to continuum strain are represented by $\check{\boldsymbol{u}}_i^V = w\langle\boldsymbol{\varepsilon}\rangle\boldsymbol{e}_i$ and the interface separation is captured in the variable $\check{\boldsymbol{u}}_i^\mathcal{I}$, such that

$$\check{\boldsymbol{u}}_i = \check{\boldsymbol{u}}_i^V + \check{\boldsymbol{u}}_i^\mathcal{I}. \tag{4.10}$$

Hence, the effective strain rate is obtained by time differentiation from Eqs. (4.9) and (4.10) and can be written in two equivalent representations:

$$\dot{\bar{\boldsymbol{\varepsilon}}} = \frac{1}{w} \sum_{i=1}^{3} \mathrm{sym}(\dot{\check{\boldsymbol{u}}}_i \otimes \boldsymbol{e}_i) = \langle\dot{\boldsymbol{\varepsilon}}\rangle + \dot{\bar{\boldsymbol{\varepsilon}}}^\mathcal{I}, \quad \dot{\bar{\boldsymbol{\varepsilon}}}^\mathcal{I} := \frac{1}{w} \sum_{i=1}^{3} \mathrm{sym}(\dot{\check{\boldsymbol{u}}}_i^\mathcal{I} \otimes \boldsymbol{e}_i). \tag{4.11}$$

Inserting this into Eq. (4.2) expands the effective stress power to

$$\bar{\mathscr{P}} = \langle\boldsymbol{\sigma}\rangle \cdot \dot{\bar{\boldsymbol{\varepsilon}}} = \langle\boldsymbol{\sigma}\rangle \cdot \langle\dot{\boldsymbol{\varepsilon}}\rangle + \langle\boldsymbol{\sigma}\rangle \cdot \dot{\bar{\boldsymbol{\varepsilon}}}^\mathcal{I}. \tag{4.12}$$

Nearby is the definition of the stress power due to continuum strains and due to interface delamination so that the following terms

$$\bar{\mathscr{P}}_V = \langle \boldsymbol{\sigma} \rangle \cdot \langle \dot{\boldsymbol{\varepsilon}} \rangle, \qquad \bar{\mathscr{P}}_{\mathcal{I}} = \langle \boldsymbol{\sigma} \rangle \cdot \dot{\bar{\boldsymbol{\varepsilon}}}^{\mathcal{I}}, \tag{4.13}$$

$$\mathscr{P}_V^{\mu} = \langle \boldsymbol{\sigma} \cdot \dot{\boldsymbol{\varepsilon}} \rangle, \qquad \mathscr{P}_{\mathcal{I}}^{\mu} = \frac{1}{V} \sum_{\xi=1}^{n^{\mathcal{I}}} \oint_{\mathcal{I}_\xi} \boldsymbol{t} \cdot \dot{\boldsymbol{d}} \, \mathrm{d}A. \tag{4.14}$$

can be identified. Together the extended Hill-Mandel condition is then denoted

$$\bar{\mathscr{P}} = \bar{\mathscr{P}}_V + \bar{\mathscr{P}}_{\mathcal{I}} = \mathscr{P}_V^{\mu} + \mathscr{P}_{\mathcal{I}}^{\mu} = \mathscr{P}^{\mu}. \tag{4.15}$$

This equation will be fulfilled in every case, when the preconditions are met, i.e., if there is no source or sink of power inside the region of interest.

Due to the fact that the interface contribution to the stress power on the microlevel $\mathscr{P}_{\mathcal{I}}^{\mu}$ is practically inaccessible in a complex three-dimensional finite element model, it can be described by the interface contribution to the stress power on the macroscopic level $\bar{\mathscr{P}}_{\mathcal{I}}$, when the classical form of the Hill-Mandel $\bar{\mathscr{P}}_V = \mathscr{P}_V^{\mu}$ condition is still valid.

Figure 4.3 is a compilation of effective stress, strain and stress power quantities, which have been described in the context of this section. They have been determined by the introduced averaging techniques, which have been applied to a unit cell with 36 grains with an impressed shear deformation of 1% (see, Figure 4.19). The details of the implemented material and interface properties will be described in the following sections. This consideration is dedicated on the gereral effective behavior of a portion of material, as described in Figure 4.1. The colors and hatching of the curves is chosen according to their meaning throughout the compilation. Figure 4.3 a) and b) show the effective shear stress component over

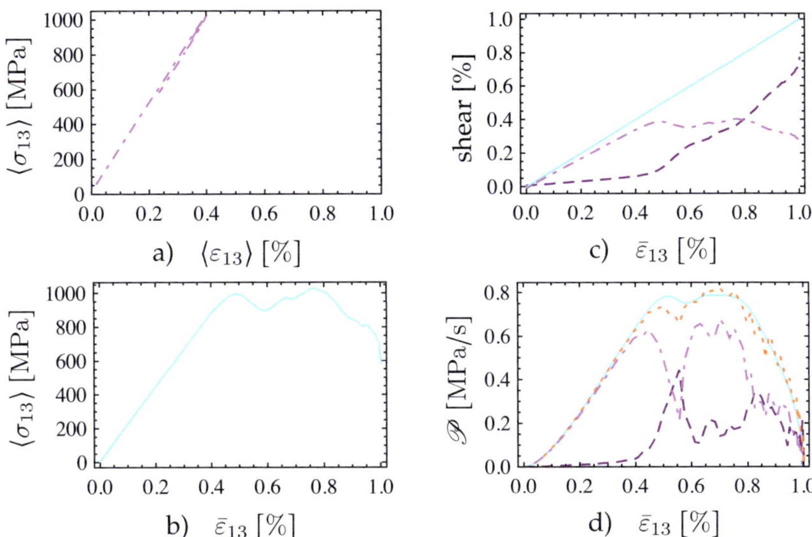

Figure 4.3: Effective quantities: Shear stress $\langle \sigma_{13} \rangle$ over a) continuum shear $\langle \varepsilon_{13} \rangle$ ($-\cdot-$), b) shear $\bar{\varepsilon}_{13}$ ($-$), c) shear quantities $\langle \varepsilon_{13} \rangle$ ($-$), $\langle \varepsilon_{13} \rangle$ ($-\cdot-$) and interface shear $\bar{\varepsilon}_{13}^{\mathcal{I}}$ ($---$), d) stress power contributions: $\bar{\mathscr{P}}$ ($-$), $\mathscr{P}_V^{\mu}$ ($-\cdot-$), $\bar{\mathscr{P}}_{\mathcal{I}}$ ($---$) and $\mathscr{P}_V^{\mu} + \bar{\mathscr{P}}_{\mathcal{I}}$ ($\cdots$),

the two different concepts of the effective strain. In a) the choosen strain is the volume average $\langle \varepsilon_{13} \rangle$, which is excluding any interface seperation in the body. In contrast in b) the axis of abscissae is the effective shear strain $\bar{\varepsilon}_{13}$ after Eq. (4.9). It can be seen, that the fracture of the material can be mainly addressed to the interface separation, whereas the continuum strain is partially reversible due to the unloading of the unit cell after the loss of the load carrying capacity. This effect is visualied in subfigure c). The three curves illustrate the concept of Eq. (4.11): The effective strain $\bar{\varepsilon}_{13}$ ($-$) can be decomposed into a continuum part $\langle \varepsilon_{13} \rangle$ ($-\cdot-$) and an integral contribution due to the interface separation $\bar{\varepsilon}_{13}^{\mathcal{I}}$ ($---$).

Subfigure d) compiles the different contributions to the extended stress power. Both the terms of Eq. (4.13) and the continuum term Eq. (4.14)$_1$ are plotted over the effective strain. It can be seen, that in the linear elastic regime the classical Hill-Mandel relation is describing the energetic behavior of the considered body in a realistic way. In the instant of fracture, this changes radically. Obviously, the continuum stress power $\mathscr{P}_V^\mu$ ($-\cdot-$) is significantly suffering from the unloading of the solids and, therefore, is beneath the effective value $\bar{\mathscr{P}}$ (—). The interface contribution $\bar{\mathscr{P}}_\mathcal{I}$ ($---$) contains the main part of the lacking portion of the stress power. The peak signalizes the fast separation at high a stress level (compare b), c) and d)). The remaining difference between the sum $\mathscr{P}_V^\mu + \bar{\mathscr{P}}_\mathcal{I}$ ($\cdots$) and the effective stress power $\bar{\mathscr{P}}$ (—) has to be attributed to dynamic effects, which are unavoidable, when a brittle material behavior is considered. After the abrupt failure, the further behavior is dominated by shear stresses due to the friction of the delaminated interfaces. In this quasi-static process, the balance of stress power after Eq. (4.15) is fulfilled.

So, it can be summarized that the energetic behavior in a body with growing cracks does not obey the classical Hill-Mandel condition, which is only considering stress power contributions due to continuum strains and stresses. Thus, an extended version of this balance equation has been suggested, which incorporates the contributions of the interface delamination. The most essential finding in terms of complex finite element simulations is that this interface contribution can be described by the consideration of quantities, which are accessible from the outer boundary and the continuum volume of the body.

4.3 Fracture Behavior of Silicon Nitride on the Local Level

4.3.1 Interface Delamination

An interface $\mathcal{I}$ can be considered as a singular surface between two solid bodies $\mathcal{B}^{\pm}$ (Figure 4.4, a)). A local coordinate system is defined by the orthogonal unit vectors $\{\boldsymbol{n}_{\mathrm{n}}, \boldsymbol{n}_{\mathrm{t}_1}, \boldsymbol{n}_{\mathrm{t}_2}\}$ according to Wei and Anand (2004). The mechanical behavior of the interface has to be described in tractions and separations, which have been defined in Section 4.2. The tractions can be decomposed into normal and tangential contributions by projections on the unit normal vectors

$$t_{\mathrm{n}} = \boldsymbol{t} \cdot \boldsymbol{n}_{\mathrm{n}}, \quad t_{\mathrm{t}_1} = \boldsymbol{t} \cdot \boldsymbol{n}_{\mathrm{t}_1} \quad \text{and} \quad t_{\mathrm{t}_2} = \boldsymbol{t} \cdot \boldsymbol{n}_{\mathrm{t}_2}. \tag{4.16}$$

With those projections,

$$\boldsymbol{t} = t_{\mathrm{n}}\boldsymbol{n}_{\mathrm{n}} + t_{\mathrm{t}_1}\boldsymbol{n}_{\mathrm{t}_1} + t_{\mathrm{t}_2}\boldsymbol{n}_{\mathrm{t}_2} \tag{4.17}$$

represents the traction vector. In perfect analogy with the tractions Eqs. (4.16) and (4.17), the separations can be split into the normal and tangential contributions.

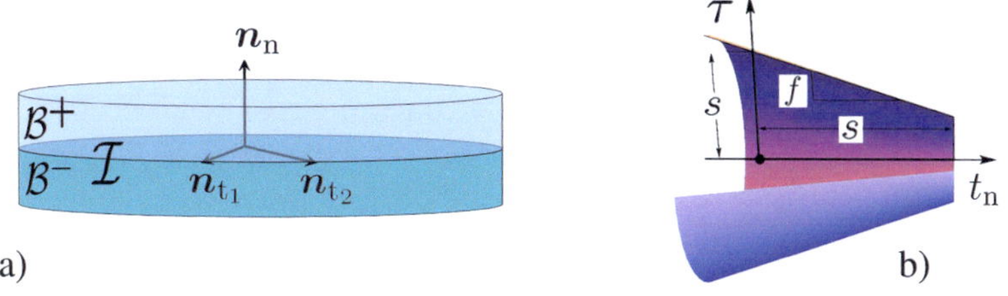

Figure 4.4: a) Schematic diagram of an interface with the local coordinate system $\{\boldsymbol{n}_{\mathrm{n}}, \boldsymbol{n}_{\mathrm{t}_1}, \boldsymbol{n}_{\mathrm{t}_2}\}$ and b) delamination conditions $\phi_{\mathrm{n}}^{\mathcal{I}} = 0$ and $\phi_{\mathrm{t}}^{\mathcal{I}} = 0$ in the t_{n}-τ-space

The delamination conditions have been assumed as an extension of Coulomb's friction law into the tensile domain. Hence, the strength and the stiffness of the interface degrade after an exceedance of the elastic admissible traction level. Furthermore, no separation is assumed to remain after unloading from any state. Figure 4.4, b) gives a graphical representation of the criteria. Due to the assumption of isotropy in the tangential plane, the traction vector in the local coordinate system can be transformed into the tangential system by

$$\boldsymbol{\tau} = t_\mathrm{n}\, \boldsymbol{n}_\mathrm{n} + \tau\, \boldsymbol{n}_\mathrm{t}, \quad \boldsymbol{n}_\mathrm{t} = \frac{1}{\tau}(t_{\mathrm{t}_1}\, \boldsymbol{n}_{\mathrm{t}_1} + t_{\mathrm{t}_2}\, \boldsymbol{n}_{\mathrm{t}_2}), \tag{4.18}$$

where τ is defined as the norm of the tangential traction components:

$$\tau = \left\| (\boldsymbol{I} - \boldsymbol{n}_\mathrm{n} \otimes \boldsymbol{n}_\mathrm{n})\, \boldsymbol{t} \right\| = \sqrt{t_{\mathrm{t}_1}^2 + t_{\mathrm{t}_2}^2}. \tag{4.19}$$

The same relationship transforms the separation vector $\boldsymbol{d}$ in the local coordinate system into the tangential space separation $\boldsymbol{\delta}$.

The conditions for tensile and shear direction can be described by

$$\phi_\mathrm{n}^{\mathcal{I}} = t_\mathrm{n} - s + q \quad \text{and} \quad \phi_\mathrm{t}^{\mathcal{I}} = \tau + f\, t_\mathrm{n} - s + q. \tag{4.20}$$

The influence of the normal traction on the shear delamination condition $\phi_\mathrm{t}^{\mathcal{I}}$ is scaled by the non-negative constant f, which is assumed to be the coefficient of friction. The initial debonding resistances for normal and tangential loading are assumed to be equal and are represented by s. The degradation stress is given by

$$q(\wp) = s(1 - \exp(-H\wp)), \tag{4.21}$$

where H is the rate of degradation and $\wp$ with $\dot{\wp} \geqslant 0$ is the internal separation-like variable due to the degradation of the interface.

The criteria can be vectorized like the tractions and separations

$$\boldsymbol{\phi}^{\mathcal{I}} := \boldsymbol{\Xi}\boldsymbol{\tau} - (s - q)\boldsymbol{i}, \quad \boldsymbol{i} = \boldsymbol{n}_\mathrm{n} + \boldsymbol{n}_\mathrm{t}, \tag{4.22}$$

where $\boldsymbol{\Xi} := \partial_{\boldsymbol{\tau}}\phi^{\mathcal{I}} = \boldsymbol{P}_{\text{nn}} + f\,\boldsymbol{P}_{\text{tn}} + \boldsymbol{P}_{\text{tt}}$ is the coupling tensor, which is identical with the traction derivative of the fracture criteria. The used projectors are defined $\boldsymbol{P}_{\text{nn}} := \boldsymbol{n}_{\text{n}} \otimes \boldsymbol{n}_{\text{n}}$, $\boldsymbol{P}_{\text{tn}} := \boldsymbol{n}_{\text{t}} \otimes \boldsymbol{n}_{\text{n}}$ and $\boldsymbol{P}_{\text{tt}} := \boldsymbol{n}_{\text{t}} \otimes \boldsymbol{n}_{\text{t}}$ with the tangential contribution implied by Eq. (4.19). It has to be mentioned, that the unit normal vectors do not depend on the traction vector $\boldsymbol{\tau}$ in the tangential space. The vectorized form simplifies the determination of the interface tangent operator, which will be carried out later in this section.

The degrading stiffness $\boldsymbol{K}$ and the compliance $\boldsymbol{S}$ of the interface are correlated by the classical relation $\boldsymbol{S} = \boldsymbol{K}^{-1}$, which induces positive definiteness. According to the choice of the delamination criteria, the stiffness and the compliance are assumed to be isotropic in the tangential plane, such that

$$\boldsymbol{K} = K_{\text{nn}}\,\boldsymbol{P}_{\text{nn}} + K_{\text{tt}}\boldsymbol{P}_{\text{tt}} \quad \text{and} \quad \boldsymbol{S} = S_{\text{nn}}\,\boldsymbol{S}_{\text{nn}} + S_{\text{tt}}\boldsymbol{P}_{\text{tt}}. \tag{4.23}$$

Following Govindjee et al. (1995), a quasi-hyperelastic framework with the notion of maximum dissipation under side conditions is used for the determination of the stress rate and the evolution of the internal variables. Hence, the Helmholtz free energy has an elastic and a degradational contribution

$$\psi = \frac{1}{2}\,\boldsymbol{\tau} \cdot \boldsymbol{S}\boldsymbol{\tau} + \mathscr{S}(\wp). \tag{4.24}$$

The degradation energy $\mathscr{S}$ depends on the accumulated degradational separation $\wp$ and can be determined from the relation $-\partial_{\wp}\mathscr{S} := q$.

With the time derivative of the Helmholtz free energy, the dissipation due to the increasing separation can be formulated. Here, the second law of thermodynamics for the isothermal case delivers $\mathscr{D} = -\dot{\psi} + \boldsymbol{\tau} \cdot \dot{\boldsymbol{\delta}} \geqslant 0$ and can be reformulated to

$$\mathscr{D} = \frac{1}{2}\,\boldsymbol{\tau} \cdot \dot{\boldsymbol{S}}\boldsymbol{\tau} + q\dot{\wp} \geqslant 0. \tag{4.25}$$

The Lagrange formula $\mathscr{L} = -\mathscr{D} + \dot{\boldsymbol{\gamma}} \cdot \boldsymbol{\phi}^{\mathcal{I}}$ with $\dot{\boldsymbol{\gamma}} = \dot{\gamma}_{n} \boldsymbol{n}_{n} + \dot{\gamma}_{t} \boldsymbol{n}_{t}$ maximizes the dissipation under the side conditions, i.e., the delamination criteria with respect to the Lagrangian multipliers $\dot{\gamma}_{i} \geqslant 0$ for the determination rates of the internal variables. An expansion of the terms delivers

$$\mathscr{L} = -\frac{1}{2}\,\boldsymbol{\tau} \cdot \dot{\boldsymbol{S}}\boldsymbol{\tau} - q\dot{\wp} + \dot{\boldsymbol{\gamma}} \cdot \left(\boldsymbol{\Xi}\boldsymbol{\tau} - (s - q)\boldsymbol{i} \right). \tag{4.26}$$

The derivatives of the Lagrange formula with respect to the stress-like variables,

$$\partial_{\boldsymbol{\tau}}\mathscr{L} \stackrel{!}{=} \boldsymbol{0} = -\dot{\boldsymbol{S}}\boldsymbol{\tau} + \boldsymbol{\Xi}\dot{\boldsymbol{\gamma}}, \tag{4.27}$$

$$\partial_{q}\mathscr{L} \stackrel{!}{=} 0 = -\dot{\wp} + \dot{\gamma}_{n} + \dot{\gamma}_{t}, \tag{4.28}$$

together with the Kuhn-Tucker conditions

$$\dot{\gamma}_{i} \geqslant 0, \quad \phi_{i} \leqslant 0 \quad \text{and} \quad \dot{\boldsymbol{\gamma}} \cdot \boldsymbol{\phi}^{\mathcal{I}} = 0 \quad \Leftrightarrow \quad \dot{\gamma}_{i}\phi_{i} = 0 \tag{4.29}$$

deliver the rate equations for the compliance tensor, when the projector properties of orthogonality and idempotency are exploited:

$$\dot{\boldsymbol{S}} = \frac{\boldsymbol{P}_{nn}\boldsymbol{\Xi}\dot{\boldsymbol{\gamma}} \otimes \boldsymbol{n}_{n}}{\boldsymbol{P}_{nn}\boldsymbol{\tau} \cdot \boldsymbol{n}_{n}} + \frac{\boldsymbol{P}_{tt}\boldsymbol{\Xi}\dot{\boldsymbol{\gamma}} \otimes \boldsymbol{n}_{t}}{\boldsymbol{P}_{tt}\boldsymbol{\tau} \cdot \boldsymbol{n}_{t}} =: \dot{S}_{nn}\,\boldsymbol{P}_{nn} + \dot{S}_{tt}\,\boldsymbol{P}_{tt}. \tag{4.30}$$

The compliance rate components for normal and shear direction have the values $\dot{S}_{nn} = (\dot{\gamma}_{n} + f\,\dot{\gamma}_{t})/t_{n}$ and $\dot{S}_{tt} = \dot{\gamma}_{t}/\tau$, respectively. For the internal degradation variable, $\dot{\wp} = \dot{\gamma}_{n} + \dot{\gamma}_{t}$ is following trivially.

The constitutive assumption for the stress-strain rate equation is

$$\dot{\boldsymbol{\tau}} = \boldsymbol{K}(\dot{\boldsymbol{\delta}} - \boldsymbol{\Xi}\dot{\boldsymbol{\gamma}}.) \tag{4.31}$$

Eq. (4.27) can be exploited, in order to obtain a different representation for the traction state:

$$\dot{\boldsymbol{\tau}} = \boldsymbol{K}(\dot{\boldsymbol{\delta}} - \dot{\boldsymbol{S}}\boldsymbol{\tau}). \tag{4.32}$$

Noteworthy is the interface tangent operator, which describes the degradation and which can be obtained from the traction rate Eq. (4.31) following Simo and Hughes (2000):

$$\dot{\boldsymbol{\tau}} = \boldsymbol{K}^{\mathcal{I}}\dot{\boldsymbol{\delta}} = \frac{\partial\dot{\boldsymbol{\tau}}}{\partial\dot{\boldsymbol{\delta}}}\bigg|_{\dot{\gamma}}\dot{\boldsymbol{\delta}} + \frac{\partial\dot{\boldsymbol{\tau}}}{\partial\dot{\gamma}}\dot{\gamma} = \left(\frac{\partial\dot{\boldsymbol{\tau}}}{\partial\dot{\boldsymbol{\delta}}} + \frac{\partial\dot{\boldsymbol{\tau}}}{\partial\dot{\gamma}}\frac{\partial\dot{\gamma}}{\partial\dot{\boldsymbol{\delta}}}\right)\dot{\boldsymbol{\delta}}. \qquad (4.33)$$

Comparing coefficients gives $\partial_{\dot{\delta}}\dot{\boldsymbol{\tau}} = \boldsymbol{K}$ and $\partial_{\dot{\gamma}}\dot{\boldsymbol{\tau}} = -\boldsymbol{K}\boldsymbol{\Xi}$. The second term of the Lagrangian multiplier contribution has to be determined from the consistency condition on the delamination criteria

$$0 = \dot{\phi}^{\mathcal{I}} = \frac{\partial\dot{\phi}^{\mathcal{I}}}{\partial\dot{\boldsymbol{\delta}}}\dot{\boldsymbol{\delta}} + \frac{\partial\dot{\phi}^{\mathcal{I}}}{\partial\dot{\gamma}}\dot{\gamma} \quad \Rightarrow \quad \frac{\mathrm{d}\dot{\gamma}}{\mathrm{d}\dot{\boldsymbol{\delta}}} = -\left(\frac{\partial\dot{\phi}^{\mathcal{I}}}{\partial\dot{\gamma}}\right)^{-1}\frac{\partial\dot{\phi}^{\mathcal{I}}}{\partial\dot{\boldsymbol{\delta}}}. \qquad (4.34)$$

The differentials can be identified by a consideration of the Eqs. (4.22) and (4.31) such that

$$\boldsymbol{J} := \frac{\partial\dot{\phi}^{\mathcal{I}}}{\partial\dot{\gamma}} = \boldsymbol{K}\boldsymbol{\Xi} + q'\,\boldsymbol{i}\otimes\boldsymbol{i} \quad \text{and} \quad \frac{\partial\dot{\phi}^{\mathcal{I}}}{\partial\dot{\gamma}} = \boldsymbol{K}\boldsymbol{\Xi}. \qquad (4.35)$$

Hence, the tangent operator can be denoted by

$$\boldsymbol{K}^{\mathcal{I}} = \boldsymbol{K}\left(\boldsymbol{I} + \boldsymbol{\Xi}\,\boldsymbol{J}^{-1}\boldsymbol{K}\boldsymbol{\Xi}\right). \qquad (4.36)$$

The allowed fracture parameters can be determined from the requirement of regularity of the tangent operator. An evaluation of the inverse determinant $\det(\boldsymbol{K}^{\mathcal{I}})^{-1} \neq 0$ delivers the upper bound

$$sH < \frac{K^0_{\mathrm{nn}}K^0_{\mathrm{tt}}}{K^0_{\mathrm{tt}} + (1-f)^2 K^0_{\mathrm{nn}}} \qquad (4.37)$$

for $\wp \to 0$, i.e., onset of delamination. The Jacobian matrix $\boldsymbol{J}$, Eq. (4.35)$_1$ yields the same information. Figure 4.5 shows the initial tangent stiffness components over the fracture parameters for a friction coefficient of $f = 0.8$ and isotropic initial stiffness ($K^0_{\mathrm{nn}}/K^0_{\mathrm{tt}} = 1$).

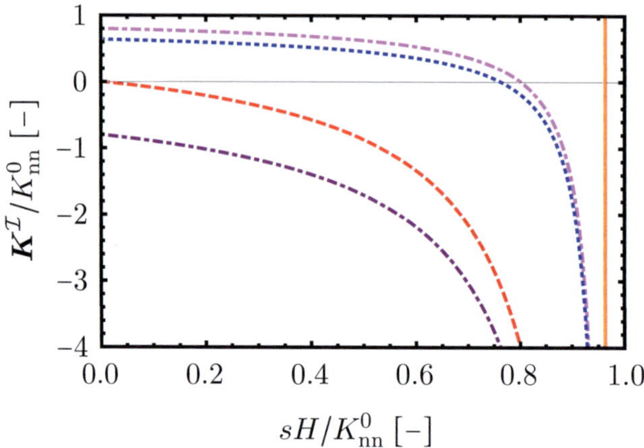

Figure 4.5: Components of the interface tangent over the delamination parameters for $K_{nn}^0 = K_{tt}^0$ and $f = 0.8$: $K_{nn}^{\mathcal{I}}$ ($--$), $K_{nt}^{\mathcal{I}}$ ($-\cdot-$), $K_{tn}^{\mathcal{I}}$ ($-\cdot-$), $K_{tt}^{\mathcal{I}}$ ($\cdots$), limit on sH (—)

It can be seen, that the pure normal component $K_{nn}^{\mathcal{I}}$ ($--$) is strictly negative, if $sH > 0$. The signature of the pure tangent component $K_{tt}^{\mathcal{I}}$ ($\cdots$) as well as the coupling component $K_{nt}^{\mathcal{I}}$ ($-\cdot-$) depend on the combination of the friction coefficient and the fracture parameters. The coupling component $K_{nt}^{\mathcal{I}}$ ($-\cdot-$) is negative in every case. The asymptotic behavior to the limit from Eq. (4.37) (—) can be seen for all curves. It has to be kept in mind, that this surprising effect of "friction hardening" is only present for $\wp \to 0$. Afterwards, the strong softening takes place at once.

The Lagrangian multipliers have to be determined by a numerical method due to the non-linearity of the problem. For the implementation into the Abaqus/Explicit user subroutine VUINTER-ACTION, the classical return-mapping algorithm after Simo and Hughes (2000) including a standard Newton-Raphson scheme on the fracture conditions has been chosen.

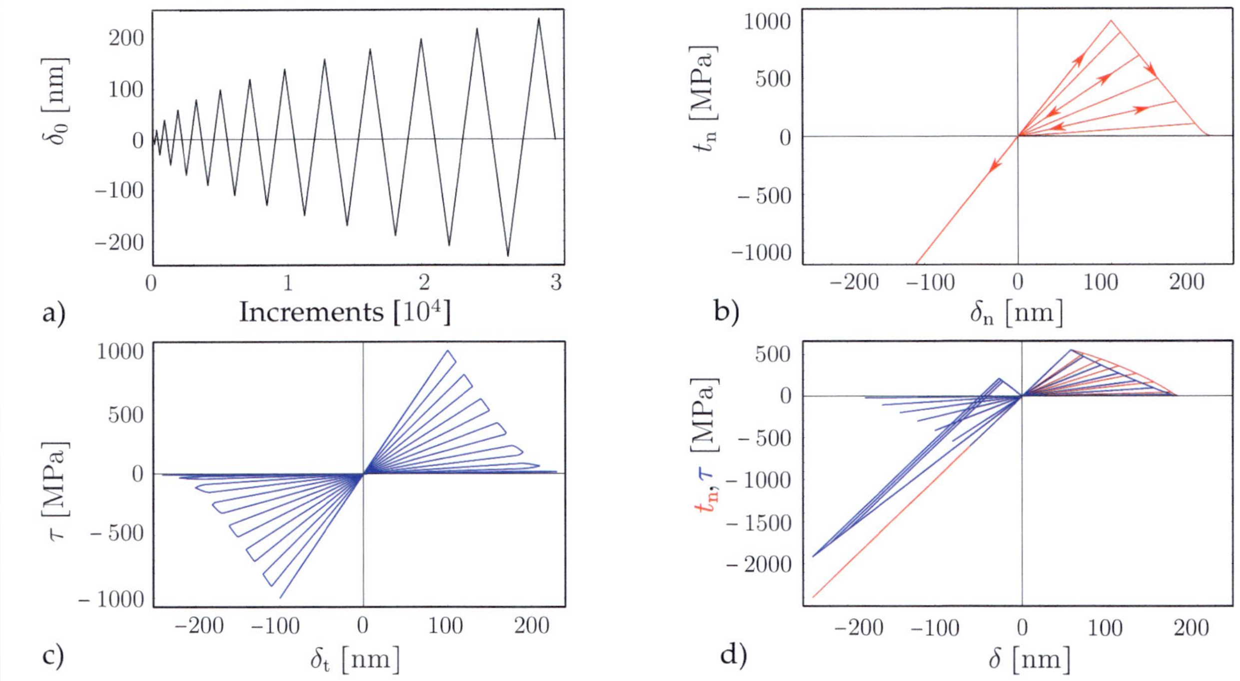

Figure 4.6: Traction-separation behavior for different load cases under the assumed load path magnitude in a), b) pure tensile loading (—), c) pure shear loading (—) and d) coupled loading (normal traction (—), shear traction (—))

In order to support the understanding of the implemented ideas, the behavior of the interface model has been depicted for different cyclic load paths, which are prescribed by the separation magnitude in Figure 4.6 a). The pure tensile traction (—) in Figure 4.6 b) shows the features of a damage behavior in tensile direction, i.e., the direct correspondence between the degradation of strength and stiffness, which leaves no remainder of inelastic separation after complete unloading of the interface. In the compressive domain, the behavior sustains its initial stiffness K_{nn}^0. The unloading cycles shows the degradation of strength and stiffness. The arrows indicate the possible loading and unloading paths, which are related to the irreversibility of the interface behavior. Figure 4.6 c) depicts the application of the magnitude in a) as shear separation (—). Here, the distinction between loading directions is not made, see, Eq. (4.20)$_2$, such that the interface degrades for negative shear tractions as well. In Figure 4.6 d) a combined load path is prescribed for both the tensile (—) and the shear direction (—). The effects of the superposition of the tractions, i.e., the friction, and their effects on the degradation behavior can be observed in detail. The softening influence or hardening effect of tensile or compressive tractions are obvious and according to the expectations.

For the unit cell fracture simulations the initial interface stiffness of $K_{nn}^0 = K_{tt}^0 = 10$ GPa/nm has been assumed. This is the highest value, which allows for a numerical stable simulation with the used framework. The initial interface strength, degradation constant and the friction coefficient have been set to $s = 1200$ MPa, $H = 0.1$ nm^{-1} and $f = 0.8$, respectively.

4.3.2 Fracture of Phases

Glassy Phase

The glassy matrix in the model is assumed to be formed by an oxynitride glass with 17% nitrogen, whose thermoelastic properties have been introduced in Section 3.3.2.

The estimation of the strength is a challenging task due to the fact, that the fracture of brittle materials is subjected to a significant volume effect. So, the Weibull theory has to be used, if the macroscopic strength results shall be used for a motivated determination from macroscopic experiments. The relation

$$\sigma_c^{\mathcal{M}}(V) = \sigma_c^{\mathcal{M}}(V_0) \left(\frac{V_0}{V} \right)^{1/m} \Gamma(1 + 1/m) \tag{4.38}$$

provides a relation between the fracture strength at different volumetric expansions (see, e.g., Gross and Seelig, 2007). Let the volume of the experimentally assessed sample be V_0. The volume of the portion, whose strength should be estimated is V. The Weibull modulus, which is a measure for the scatter in the experimental data is denoted by m. As macroscopic strength values, the experimental data from Iba et al. (1999) is used. Here, fibers with a diameter $d_0 = 20\,\mu$m were made from oxynitride glass. These fibers resisted stresses between 3 and 5 GPa. Together with an assumed fiber length of $l_0 = 30\,$mm the fiber volume is $V_0 = \pi/4\, d_0^2\, l_0 = 9.42{\cdot}10^{-3}\,$mm^3. For the Weibull modulus of glass fibers Le Bourhis (2008) provides a wide range between 10 and 30. An estimation of a glass pocket is possible, if a larger glass pocket in Figure 2.4 is considered. If it is assumed to have the shape of a regular tetrahedron with an edge length of 1 μm, then the volume would be $V = l^3/4\sqrt{3} = 1.44{\cdot}10^{-10}\,$mm^3. If those assumptions are put together, the strength map in Figure 4.7 a) can be obtained from Eq. (4.38). It can be seen that a strong dependence

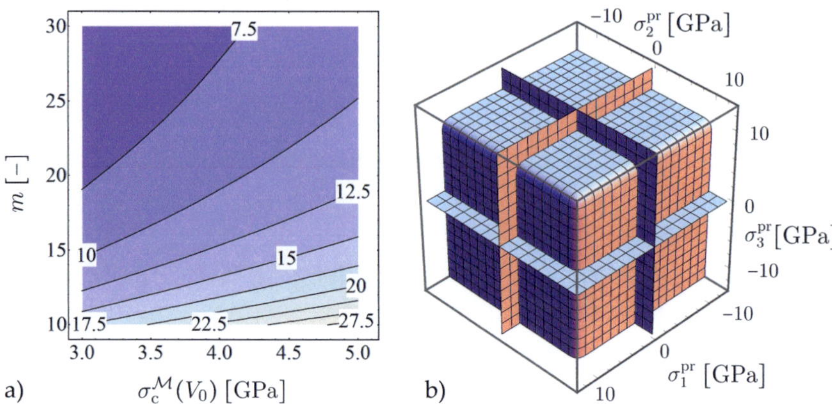

Figure 4.7: Fracture stress of the glassy phase: a) Contour plot of fracture stress of the glassy phase $\sigma_c^{\mathcal{M}}$ [GPa] based on the assumptions for Weibull modulus m and glass fiber strength $\sigma_c^{\mathcal{M}}$, b) Contour plot of the isotropic failure criterion $\phi^{\mathcal{M}} = 0$ for the glassy matrix phase in the principal stress space, $\sigma_c^{\mathcal{M}} = 10$ GPa, $p = 20$; planes indicate vanishing stresses.

is given especially for the Weibull modulus. The predicted strength values reach from $\sigma_c^{\mathcal{M}} = 5.36\ldots28.7$ GPa. Hence, the assumption of $\sigma_c^{\mathcal{M}} = 10$ GPa appears as reasonable choice. The fracture behavior of the glassy matrix phase $\mathcal{M}$ is assumed to be caused by the maximum tensile stress. They are obtained from the eigenvalue problem $\boldsymbol{\sigma} \boldsymbol{p}^{\mathrm{pr}} = \sigma^{\mathrm{pr}} \boldsymbol{p}^{\mathrm{pr}}$ with the principal stresses σ^{pr} and the corresponding eigenvectors $\boldsymbol{p}^{\mathrm{pr}}$. The problem is solved numerically by a Jacobi-scheme (see, e.g., Nipp and Stoffer, 2002). The fracture criterion is formulated as ℓ^p-norm on the tensile part of the principal stresses, where $1 \leqslant N \leqslant 3$ is the number of different eigenvalues of the stress tensor $\boldsymbol{\sigma}$ and $\sigma_c^{\mathcal{M}}$ is the cracking stress of the matrix material. Thus, the criterion can be denoted

$$\phi^{\mathcal{M}} = \left(\sum_{\xi=1}^{N} \left(\max\left\{0, \sigma_\xi^{\mathrm{pr}}\right\} / \sigma_c^{\mathcal{M}} \right)^p \right)^{1/p} - 1. \tag{4.39}$$

The fracture criterion transforms into the strict maximum principal stress criterion, if $p \to \infty$. In Figure 4.7 b) the contours the fracture criterion $\phi^{\mathcal{M}} = 0$ in the principal stress space are depicted. Here, the assumed cracking stress is 10 GPa.

β-Si$_3$N$_4$ Grains

The cracking criterion for the β-phase $\mathcal{G}$ is motivated by the transverse-isotropy of the elastic stiffness tensor. Hence, the positive eigenvalues $\max\{0, \sigma_\xi^{\mathrm{pr}}\}$ and the corresponding projectors $P_\xi^{\mathrm{pr}} := p_\xi^{\mathrm{pr}} \otimes p_\xi^{\mathrm{pr}}$ have been used for the tensile stress tensor in spectral representation $\sigma^+ = \sum_{\xi=1}^{N} \max\{0, \sigma_\xi^{\mathrm{pr}}\}\, P_\xi^{\mathrm{pr}}$.

The projectors $P_\zeta^{\mathcal{G}} = n_\zeta^{\mathcal{G}} \otimes n_\zeta^{\mathcal{G}}$, $\zeta = 1 \ldots 4$ have been constructed from the normal directions of a grain (Section 2.2.2). The tensile stress tensor is mapped on those projectors in order to capture the anisotropy. The ℓ^p-norm for the grain $\phi^{\mathcal{G}}$ fracture criterion takes the form

$$\phi^{\mathcal{G}} = \left(\sum_{\zeta=1}^{4} \left| P_\zeta^{\mathcal{G}} \cdot \sigma^+ / \sigma_{c,\zeta}^{\mathcal{G}} \right|^p \right)^{1/p} - 1. \tag{4.40}$$

Figure 4.8 a) shows the three-dimensional contour of the fracture criterion $\phi^{\mathcal{G}} = 0$ for the grain, depicted on the right. Motivated by the ab-initio simulation results of Ogata et al. (2004), the selected parameters are $\sigma_{c,1\ldots3}^{\mathcal{G}} = 1923$ MPa and $\sigma_{c,4}^{\mathcal{G}} = 2000$ MPa and $p = 20$. The anisotropy and the influence of the crystal planes is slightly pronounced. In axial direction, the fracture stress is approximately 4% higher than the fracture stress in the basal direction. In the basal plane, a slight influence of the used normal vectors is visible. In Figure 4.8 b) the influence of the orientation on the cracking direction is visualized. The notched sample is thought to be cut out of a single crystal. The directions are indicated by the small coordinate system, which is drawn in to the sketches in the lower row of the compilation. The sample is loaded by displacements on the front faces of the sample, as it is indicated by the vertical arrows. The upper row shows the distribution of the fracture criterion $\phi^{\mathcal{G}}$. Green dyeing indicates high stress levels around the notch tip shortly before cracking.

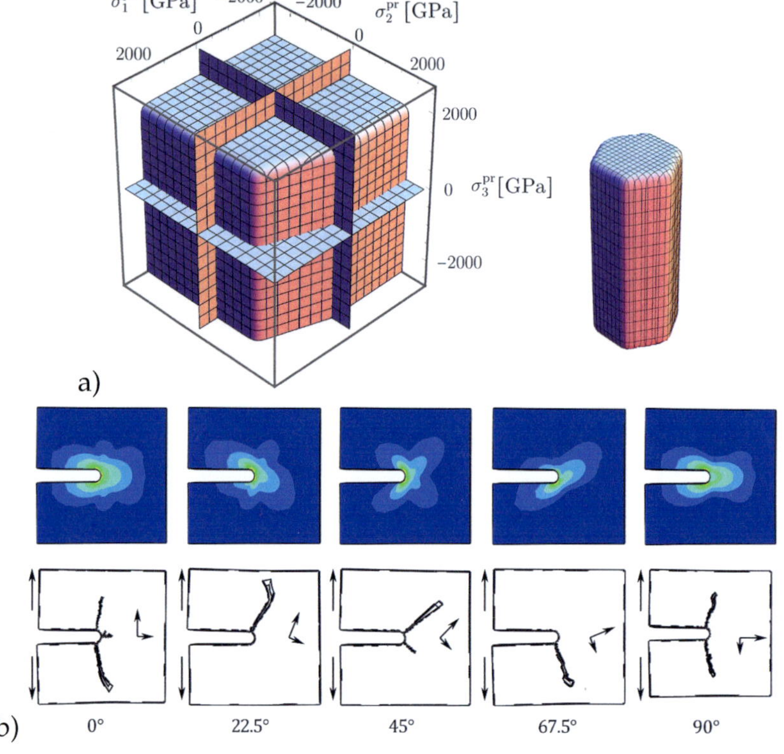

Figure 4.8: a) Contour plot of the fracture criterion for the β-grain $\phi^{\mathcal{G}} = 0$ in the principal stress space; planes indicate vanishing stresses; b) examination of the fracture behavior on a simplified compact tension specimen with different grain orientations

It can be seen that the orientation of the contour lines corresponds with the crystal orientation. The lower row shows the cracks. Interesting is the fact that the sample never breaks finally in the direction of the notch, although an onset of cracking occurs in case of the crystal, which is oriented in loading direction. In general, a cracking into the direction of the weaker stress can be observed in all cases.

The fracture criteria are implemented as user subroutine VUMAT for Abaqus/Explicit. In integration points, where the fracture criteria $\phi^{M/\mathcal{G}}$ are fulfilled, stress is set to zero.

4.4 Fracture Behavior of the Effective Material

The fracture of the effective material has been assumed to occur under maximum principal stresses. This has been implemented into a continuum damage mechanics model. The criterion is denoted

$$\bar{\phi} = \max_{\xi=1}^{N}\{\bar{\sigma}_{\xi}^{\mathrm{pr}}\} - \bar{\sigma}_{\mathrm{c}} + \bar{q}, \tag{4.41}$$

where $\bar{q}$ is given in analogy to Eq. (4.21), redefining all parameters by effective parameters for the continuum model. The fracture model uses the same thermodynamical consistent framework like the interface delamination behavior. As already described in Section 4.3.1, the material is assumed to remain elastic after it undergoes damage, i.e., degradation of strength and stiffness. Consequently, following again Govindjee et al. (1995), the Helmholtz free energy for the quasi-hyperelastic continuum has the form

$$\bar{\psi} = \frac{1}{2}\bar{\sigma} \cdot \bar{\mathbb{S}}[\bar{\sigma}] + \bar{\mathscr{S}}(\bar{\wp}). \tag{4.42}$$

The effective degradational energy $\bar{\mathscr{S}}(\bar{\wp})$ is defined in analogy to the corresponding quantity for the interface degradation, introduced in Eq. (4.21). The compliance tensor and the stiffness tensor are linked by the relation $\mathbb{S} = \mathbb{C}^{-1}$. The dissipation obtains the form

$$\bar{\mathcal{D}} = \frac{1}{2}\bar{\boldsymbol{\sigma}} \cdot \dot{\mathbb{S}}[\bar{\boldsymbol{\sigma}}] + \bar{q}\dot{\wp} \geqslant 0, \tag{4.43}$$

such that the Lagrange formula for the determination of the internal variables is

$$\bar{\mathcal{L}} = -\frac{1}{2}\bar{\boldsymbol{\sigma}} \cdot \dot{\mathbb{S}}[\bar{\boldsymbol{\sigma}}] - \bar{q}\dot{\wp} + \dot{\bar{\gamma}}\left(\max_{\xi=1}^{N}\{\bar{\sigma}_{\xi}^{\mathrm{pr}}\} - \bar{\sigma}_{\mathrm{c}} + \bar{q} \right). \tag{4.44}$$

An evaluation of the conditions for the maximum dissipation

$$\partial_{\bar{\boldsymbol{\sigma}}}\bar{\mathcal{L}} \overset{!}{=} \mathbf{0} = -\dot{\mathbb{S}}[\bar{\boldsymbol{\sigma}}] + \dot{\bar{\gamma}}\,\bar{\boldsymbol{P}}, \quad \bar{\boldsymbol{P}} := \partial_{\bar{\boldsymbol{\sigma}}}\bar{\phi} = \bar{\boldsymbol{p}}_{\mathrm{max}}^{\mathrm{pr}} \otimes \bar{\boldsymbol{p}}_{\mathrm{max}}^{\mathrm{pr}}, \tag{4.45}$$

$$\partial_{\bar{q}}\bar{\mathcal{L}} \overset{!}{=} 0 = -\dot{\wp} + \dot{\bar{\gamma}} \tag{4.46}$$

delivers the rates of the internal variables. The form of the fracture criterion allows an application of the linear map for the derivation of the compliance rate in the sense of an associated flow rule

$$\dot{\mathbb{S}} = \dot{\bar{\gamma}}\,\frac{\bar{\boldsymbol{P}} \otimes \bar{\boldsymbol{P}}}{\bar{\boldsymbol{\sigma}} \cdot \bar{\boldsymbol{P}}}. \tag{4.47}$$

Here, the dissipation relevant direction, i.e., the direction of the maximum principal stress is considered only. The rate of the strain-like damage variable follows trivially $\dot{\wp} = \dot{\bar{\gamma}}$.

The stresses can be determined from the rate form of Hooke's law

$$\dot{\bar{\boldsymbol{\sigma}}} = \bar{\mathbb{C}}[\dot{\bar{\boldsymbol{\varepsilon}}} - \dot{\bar{\gamma}}\bar{\boldsymbol{P}}]. \tag{4.48}$$

It has to be mentioned that this definition of the stress-rate is symmetric with respect to tensile and compressive loads, which is cumbersome, if crack closure plays a role for the material behavior, e.g., for cyclic loading. An option for a distinction between the tensile or the compressive domain can be obtained by the consideration of the

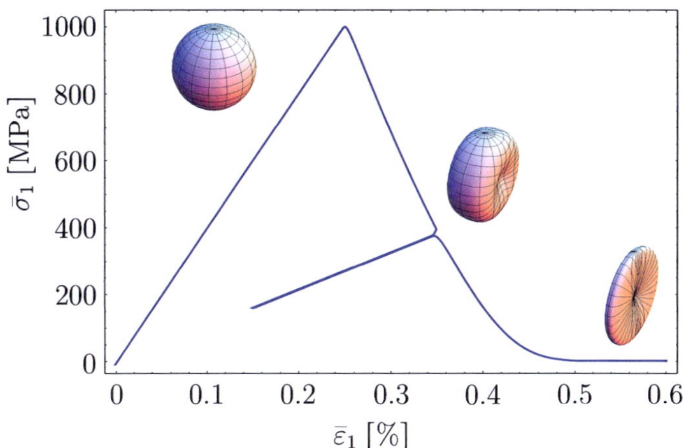

Figure 4.9: Stress-strain curve for continuum undergoing elastic damage; contours describe the degradation of the effective elastic stiffness $\bar{\mathbb{C}}$.

projection of the present strain tensor $\bar{\varepsilon}$ on normal direction of the crack plane $\bar{P}$. Tensile or orthogonal modes deliver $\bar{\varepsilon} \cdot \bar{P} \geqslant 0$ and compressive loading is characterized by $\bar{\varepsilon} \cdot \bar{P} < 0$. For the case of crack closure, the initial stiffness is a reasonable choice for the stiffness of the material. Nevertheless, such discontinuity would give rise to certain issues on the definition of the energy formulation. Due to the fact that for the present cases, the crack closure case is not considered, this questions are not examined in more detail here.

Like in the case of the interface behavior, the Lagrange multiplier is determined numerically by the standard procedure after Simo and Hughes (2000).

Figure 4.9 shows a typical stress strain curve for a uniaxial tensile deformation of the material in e_1-direction. The constants for the initial elastic stiffness are taken from Section 3.5. The effective fracture strength is $\bar{\sigma}_c = 1000\,\mathrm{MPa}$ and the effective degradation constant is $\bar{H} = 200$. The curve shows, how the continuum reaches the point of material instability. Afterwards, the resistance and the stiffness degrade, such that after unloading no inelastic strain remains, which is symbolized by the unloading cycle. The spherical projections of the elastic stiffness (Böhlke and Brüggemann, 2001) show the evolving transverse-isotropy of the stiffness tensor $\bar{\mathbb{C}}$, due to the degradation process.

For the issues of uniqueness and well-posedness of problems in continuum mechanics a consideration of the tangent operator for damage $\bar{\mathbb{C}}^{\mathrm{d}}$ is important. It is given by the relationship

$$\bar{\mathbb{C}}^{\mathrm{d}} = \frac{\mathrm{d}\bar{\boldsymbol{\sigma}}}{\mathrm{d}\bar{\boldsymbol{\varepsilon}}}. \tag{4.49}$$

The determination of this total differential is based on the consideration of the total differentials of stress Eq. (4.48) and the cracking condition Eq. (4.41) with respect to the strain and delivers (Simo and Hughes, 2000)

$$\bar{\mathbb{C}}^{\mathrm{d}} = \bar{\mathbb{C}} - \frac{\bar{\mathbb{C}}[\bar{\boldsymbol{P}}] \otimes \bar{\mathbb{C}}[\bar{\boldsymbol{P}}]}{\bar{\boldsymbol{P}} \cdot \bar{\mathbb{C}}[\bar{\boldsymbol{P}}] - \bar{q}'}, \quad \bar{q}' = \bar{H}\bar{\sigma}_{\mathrm{c}} \exp(-\bar{H}\bar{\wp}). \tag{4.50}$$

Care has to be taken for the choice of the softening constants, because a vanishing denominator has to be avoided. At the onset of cracking, i.e., $\bar{\wp} = 0$, the initial softening modulus is $q_0' = \bar{H}\bar{\sigma}_{\mathrm{c}}$. Hence, it is worth to see, if there is an absolute limit for the softening constants. The denominator can be simplified further by the evaluation of the projection of the isotropic stiffness $\bar{\mathbb{C}} = \bar{\lambda}\,\boldsymbol{I} \otimes \boldsymbol{I} + 2\bar{G}\,\mathbb{I}^{\mathrm{S}}$ on the cracking direction $\boldsymbol{P}$:

$$q_0' < \min_{\bar{\boldsymbol{P}}}(\bar{\boldsymbol{P}} \cdot \mathbb{C}[\bar{\boldsymbol{P}}]) = \bar{\lambda}\,\min_{\bar{\boldsymbol{P}}}(\mathrm{tr}^2(\bar{\boldsymbol{P}})) + 2G. \tag{4.51}$$

Because of $\mathrm{tr}^2(\bar{\boldsymbol{P}}) \geqslant 0$, the upper limit for the initial derivative is

$$\bar{q}_0' = \bar{\sigma}_{\mathrm{c}}\bar{H} < \lambda\,\mathrm{tr}^2(\bar{\boldsymbol{P}}) + 2\bar{G}. \tag{4.52}$$

Hence, the choice of the parameters is case sensitive. A pure tensile stress field allows a higher degradation rate, than shear stress.

The spontaneous loss of definiteness in the moment, when fracture occurs, can be shown by the scalar product $\bar{\boldsymbol{P}} \cdot \bar{\mathbb{C}}^{\mathrm{d}}[\bar{\boldsymbol{P}}]$ with any cracking direction tensor $\bar{\boldsymbol{P}} \neq \boldsymbol{0}$. An application of the linear map delivers

$$\bar{\boldsymbol{P}} \cdot \bar{\mathbb{C}}^{\mathrm{d}}[\bar{\boldsymbol{P}}] = -\frac{\bar{q}'\,\bar{\boldsymbol{P}} \cdot \bar{\mathbb{C}}[\bar{\boldsymbol{P}}]}{\bar{\boldsymbol{P}} \cdot \bar{\mathbb{C}}[\bar{\boldsymbol{P}}] - \bar{q}'}, \tag{4.53}$$

which is strictly negative for any admissible initial softening constants. Consequently, mechanical problems, which use this constitutive assumption have to be regularized.

4.5 Results

4.5.1 Unit Cell Approach for the Determination of the Effective Fracture Behavior

The micromechanical finite element model has to incorporate all described constitutive and geometric features as well as the boundary conditions. The generated geometry (see, Chapter 2) is meshed with ScanIP from Simpleware (Young et al., 2008). In Figure 4.10 a schematic of a periodic unit cell with the phase- and interface-wise constitutive assumptions is depicted. The used symbols correspond with the symbols in the previous sections. The material orientations are delivered by the structure generator.

The teal layer ($\bullet$) is necessary to avoid a strong influence of small imperfections to periodicity of the unit cells after the meshing with ScanIP, which arise from the smoothing technique. So, it occurs that grain interfaces on opposing sides of the unit cube are not matching perfectly to each other, which would cause significant artifacts.

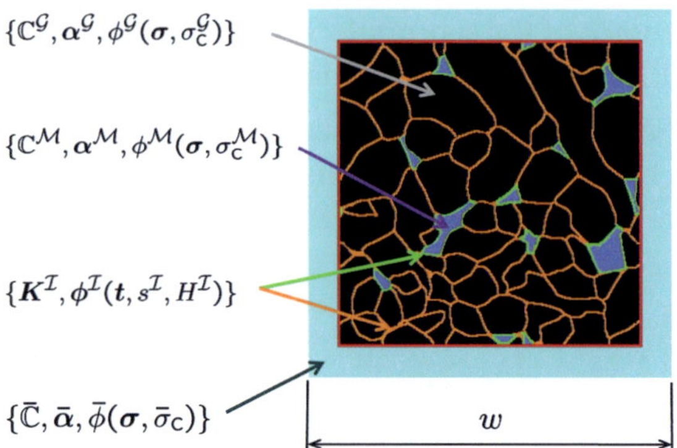

Figure 4.10: Schematic of a periodic unit cell embedded into effective material ($\bullet$) with grains ($\mathcal{G}$), glassy matrix ($\mathcal{M}$) and interfaces ($\mathcal{I}$) between grains and grains (—) and matrix (—)

A feasible workaround is the application of the projected periodic boundary conditions (Section 3.4.2) on a layer with the effective thermoelastic properties (Section 3.5). The resulting boundary conditions are a linear combination of the imposed displacement field and the surface tractions implied by those displacements through the elastic properties of the layer. More generally this is a weighted combination of the Dirichlet and Neumann type boundary conditions. This third type of boundary conditions is often called after Robin, but other eponyms are possible as well (Gustafson and Abe, 1998).

The consideration of the degrading material behavior leads in general to a local loss of Hadamard's strong ellipticity condition (see, e.g., Bažant and Belytschko, 1985; Sluys, 1992; Govindjee et al., 1995). The result is an ill-posed problem, which makes a regularization inevitable. A well introduced option for a regularization of the static problem is the solution of a related dynamic problem, such that the inertial forces due to mass density and material velocity have to be incorporated into the balance equation. In order to obtain a quasi-static solution, the dynamic effects from the inertial contributions must not influence the solution of the considered problem. Here, an estimation of reasonable values of the imposed deformation speed and the speed of sound in the material is required. Following Kinsler et al. (2000), the wave speeds in an isotropic and homogeneous solid aggregate are determined by the bulk and shear modulus and its density. So, the longitudinal and shear wave speed is given by $c_1^2 = (K + 4/3\,G)/\rho$ and $c_s^2 = G/\rho$, respectively. With the values for the effective properties from Table 3.1 at room temperature and a density of bulk silicon nitride of $\rho = 3.2$ kg/l the tensile and shear relevant velocity are $c_1 = 11.1$ km/s and $c_s = 6.1$ km/s.

A quasi-static deformation of a homogeneous unit cube with an initial edge length w is obtained, if the deformation time $t_{\bar{u}} = \bar{u}/\bar{v} = \bar{\varepsilon}w/\bar{v}$ is several orders of magnitude smaller than the wave travel time $t_{c_{1/s}} = w/c_{1/s}$ in the cube for the considered deformation mode. Hence, an estimation of the time or speed ratio can be given by $t_{\bar{u}} \leqslant \xi\, t_{c_{1/s}}$ or $\bar{v} \leqslant \xi\,\bar{\varepsilon}\,c_{1/s}$ with $\xi \ll 1$. An appropriate selection of the factor ξ is crucial for the attainment of a realistic determination of a material behavior. Reasonable choices are between 0.1 and 1% of the relevant wave travel time. The displacement field has to be applied by a smooth technique to avoid velocity and acceleration jumps, which would induce uninteded shock waves in the model and with it dynamic artifacts. This means, the velocity and acceleration has to vanish at the beginning of the deformation process.

4.5.2 Structure-Property Relation

The microstructure of silicon nitride together with the grain boundary toughness are the crucial properties for the design of a well-balanced fracture strength and toughness. These properties can be assessed on the macroscopic level by *R*-curve measurements. The curves describe the relation between fracture toughness and crack extension as it is outlined in detail in Munz (07). In general, a rising R-curve is indicating a tough material. In silicon nitride, four stages of the R-curve can be distinguished, which is explained in Fünfschilling et al. (2011). The first domain characterizes the fracture behavior during the first micrometers of the crack extension. Here, the microscale mechanisms, i.e., the delamination and cracking of the large and elongated β-Si_3N_4 grains are the dominating effects. The most important finding is that the steep increase in toughness can be addressed to elastic bridges, which are formed by large grains. Those grains are perfectly bound to other grains and/or to the glassy phase, while a crack is propagating around them. Depending on the angle of incidence and the grain boundary toughness, the crack is circumventing the large grain and grows forward through smaller grains. This results in elastic bridges, which are formed by the large grains. These are the main reinforcement contribution in the early crack stage and, therefore, improve the fracture toughness significantly. The highly heterogeneous stresses, which are carried by the bridging grains are regarded as bridging stresses. In the classical fracture mechanics theory, an effective bridging stress field can be determined from the R-curve by the usage of weight functions (Fett et al., 2005). Differently to the R-curves, the bridging stress fields are seen as material property, which is induced by the microstructure and which, therefore, remains uninfluenced by the geometry of the considered specimen (Fünfschilling et al., 2011). This is essential for an experimental and macroscopic characterization of a structural-reinforced ceramics material. For a microscopic assessment, based on a space-resolved finite element simulation framework, a detailed consideration would be premature, such that this research direction is not followed further in the context of this work.

Due to the complex experimental assessment, the first part of the crack resistance curve was firstly documented by Fett et al. (2008),

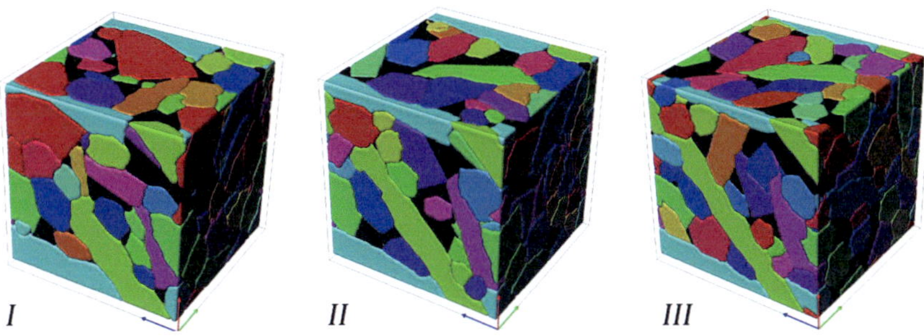

Figure 4.11: Three unit cells with different mean aspect ratios: cell I ($\langle \mathcal{A}^I \rangle = 3.07$), cell II ($\langle \mathcal{A}^{II} \rangle = 4.03$) and cell III ($\langle \mathcal{A}^{III} \rangle = 4.93$)

where a special four-point bending test on notched beams together with an elaborated semi-analytical evaluation technique was used to distinguish between the influences of the notch radius and the short cracks. Here, the finding that silicon nitride gains 90% of the its final fracture toughness during the first 10 μm of the crack propagation.

In order to give further insight into the acting mechanisms, this section will examine the influence of different grain-length to aspect ratio distributions at a given grain boundary toughness under uniaxial tensile deformations with effective free transverse contraction.

For this examination, three different periodic unit cells in the further context with 64 grains, designated with I, II and III have been created. The cells with mean aspect ratios of $\langle \mathcal{A}^I \rangle = 3.07$, $\langle \mathcal{A}^{II} \rangle = 4.03$ and $\langle \mathcal{A}^{III} \rangle = 4.93$ have been generated in eight seeding steps with eight grains each. The width w of the cells is 3 μm. The cells can be seen in Figure 4.11. The difference in the mean grain shapes appears in all cases.

The usage of the same random seed for the structure generation algorithm allows for a comparison of the different realizations of the unit cell. Only the axial growth velocity has been varied.

The cells are deformed uniaxially with a free transverse contraction by the strain tensor $\bar{\varepsilon} = \bar{\varepsilon}_0(\boldsymbol{B}_i + \bar{\nu}(\boldsymbol{I} - \boldsymbol{B}_i))$, $i = 1\ldots3$. The following images in Figure 4.12 and 4.13 show the degradation process for the unit cell *III* with the highest mean aspect ratio, when deformed along the e_3-axis. On the left side of the images, the displacement field in the loading direction u_3 is shown. The right side is reserved for the corresponding stress fields σ_{33}.

Figure 4.12 a) shows the state after the cooling down from the sintering temperature. Thus, only thermal strains and residual stresses are present. The displacement field is smooth and characterized by slight fluctuations due to the phase contrasts. The subfigure b) shows the state, where the effective mechanical strain in loading direction is equilibrating the thermal strain. This leads to considerable stresses in the microstructure. A load transition to the large grain, which is mostly aligned with the e_3-axis, can be observed. Subfigure c) shows the onset of delamination, which is visible due to a first displacement jump. The images in Figure 4.13 shows the progress of the final fracture. It can be seen, that the bridging grain is carrying a large amount of the load. The displacement field now shows a new jump on the right side of the model. Subfigure b) and c) visualize the fracture of the bridging grain, which leads to material instability.

The findings from the image series can be summarized as follows: The main features of the fracture in silicon nitride are represented in the model in a simplified, but nevertheless significant way. The crack path deviation due to elastic bridging by long and elongated grains is clearly visible.

In order to examine the influence of the structural properties on the effective behavior a consideration of the relations between effective stress $\langle\sigma\rangle$ and strain $\bar{\varepsilon}$ in tensile direction have been considered, see, Eqs. (4.4) and (4.9). Those two quantities allow for the determination of the fracture toughness, when certain assumptions are made.

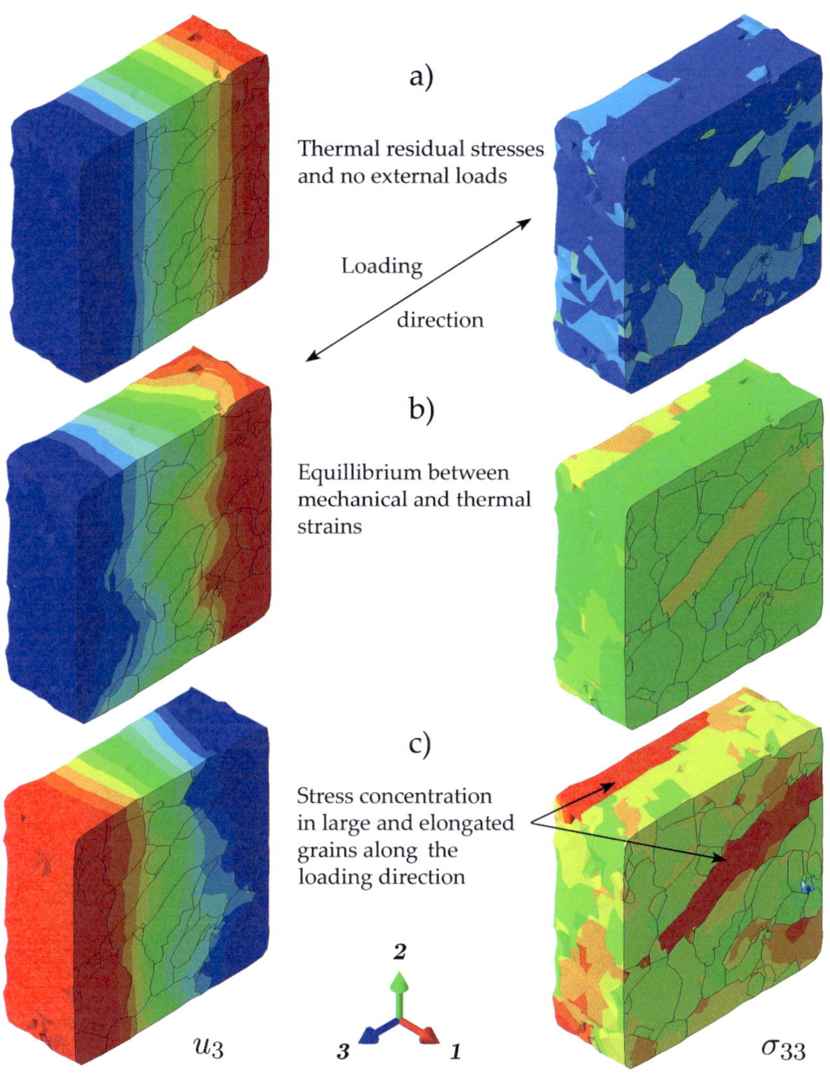

Figure 4.12: Time evolution of fracture for unit cell $III - 1$

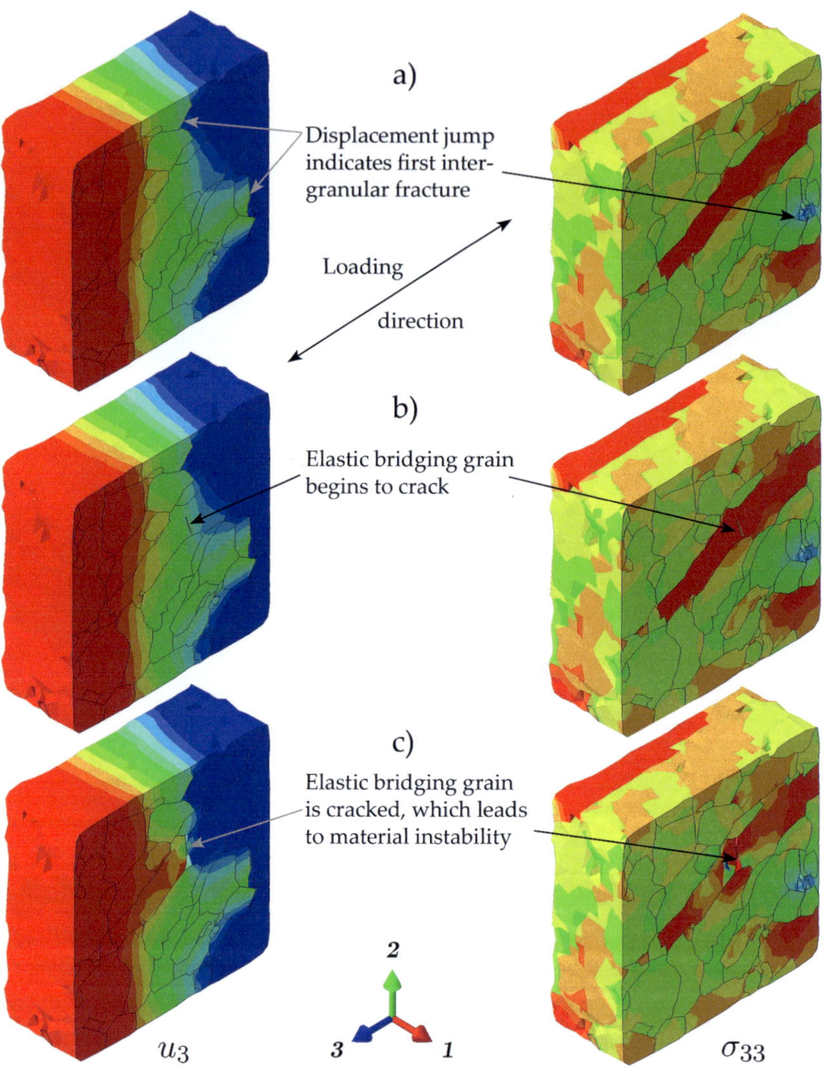

Figure 4.13: Time evolution of fracture for unit cell $III - 2$

The fracture toughness of a material is characterized by the energy release rate $\mathscr{G}$, i.e., the variation of the internal energy due to the crack propagation. Following the basic relations of the linear fracture mechanics, as it is outlined in detail in Munz and Fett (1999), the energy release rate is given by

$$\mathscr{G} = \frac{1}{2} F^2 \frac{\mathrm{d}S}{\mathrm{d}A}. \tag{4.54}$$

Here, F is assumed to be the force on the surface of the unit cube, such that $F := \langle \sigma \rangle\, w^2$. The compliance is denoted by $S = u/F$ with the boundary displacement $u = \bar{\varepsilon} w$. The cracked area A has to be determined under the assumption of a certain fracture mode, because here information of the crack geometry is needed. Therefore, the stress intensity factor for fracture under mode I is considered as relevant for the uniaxial tensile load of a brittle material. This factor has two representations, which have to be exploited. First, it is a relation between the critical stress $\langle \sigma \rangle$ or the related force F and the crack length a:

$$K_{\mathrm{R}} = \langle \sigma \rangle \sqrt{a} Y = \frac{F}{w^2} \sqrt{a} Y. \tag{4.55}$$

The geometry factor Y is the geometry function, which plays a crucial role for the evaluation of fracture experiments. For a basic understanding, it can be considered as a constant for a certain crack shape and, therefore, a value of 1 is assumed. The second relation exists between the energy release rate and the stress intensity factor. Given the effective uniaxial stress state, which is a special case of plane stress, the relation

$$K_{\mathrm{R}}^2 = \mathscr{G}\, \bar{E}, \tag{4.56}$$

holds, where $\bar{E}$ is initial Young's modulus. Introducing the Eqs. (4.54) and (4.55) into Eq. (4.56) delivers

$$\mathrm{d}A = \frac{w^4 \bar{E}}{2\sqrt{A}\, Y^2}\, \mathrm{d}S, \tag{4.57}$$

when the crack area is assumed have a unit shape. From this relation, an incremental calculation of the crack surface is possible.

In Figure 4.14 the effective stress-strain curve, the compliance, the energy release rate and the stress intensity factor are compiled for the different geometries and for representative load paths. Figure 4.14 a) shows the different tensile behaviors. A significant improvement of the fracture strength can be seen. This trend can be attributed to the higher content of large and elongated grains, which are transferring higher local loads. In Figure 4.14 b) the compliance S^* is depicted for the fracture domain over the effective strain. In order to achieve a better feeling for the quantities, the curves are normalized on the initial Young's modulus $\bar{E}$ and on the edge length w of the unit cell, so that $S^* = ES\bar{w} = \bar{E}\bar{\varepsilon}/\langle\sigma\rangle$. The curves start, where the degradation of the material, mainly due to interface delamination begins. In the moment of fracture compliance rises quickly. The influence of the structural reinforcement can be seen.

Of special interest are the subfigures c) and d). Both curves are normalized on the unit cell size w. The most obvious feature is the general rising trend of the curves, which corresponds to experimental observations of Fett et al. (2008). Important appears the feature that here a significant influence of the grain shape can be observed. The clear trend: The higher the aspect ratio, the higher and steeper is the R-curve.

It is naturally that an effect of the loading direction can be observed. This is, basically, due to the limited size of the considered unit cells. However, the general trend towards increasing fracture strength and resistance with increasing aspect ratios is observable, as it can be seen in Figure 4.15. The influence of the geometry can be seen in all cases and becomes more pronounced with increasing aspect ratio. The microscopic reasons for the differences in the macroscopic behaviors can be observed in Figure 4.16. Here, the corresponding

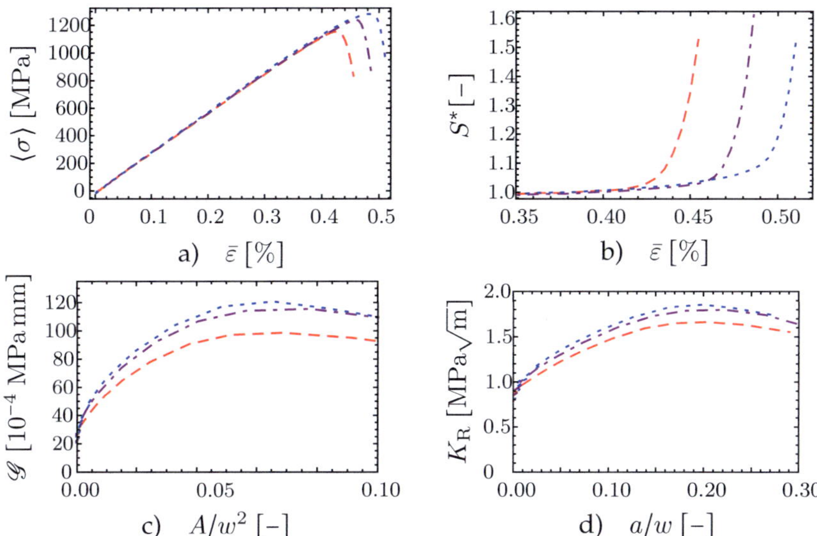

Figure 4.14: Influence of the mean aspect ratio on different effective properties; $\langle \mathcal{A}^{I} \rangle$ $(- -)$, $\langle \mathcal{A}^{II} \rangle$ $(- \cdot -)$, $\langle \mathcal{A}^{III} \rangle$ $(\cdots)$; a) effective stress-strain curve, b) normalized compliance S^* over the effective strain, c) energy release rate $\mathcal{G}$ over crack surface (normalized on the unit cell cross section w^2) and d) stress intensity factor K_{R} over crack length (normalized on the edge length of the unit cell w)

displacement (left) and stress fields (right) for the microstructure *II* with the medium mean aspect ratio is gathered for the two loading directions with the biggest difference in the macroscopic behavior. The values of the displacement field are minimum and maximum values at the present state. The stress field color code is starting with blue color (•) at −100 MPa, which represents the compressive stresses in the grains due to thermal strains and is ending in red color (•) at 2000 MPa, which is the fracture strength of the grains in axial direction. The high values of strength and fracture toughness, which have been measured for the loading direction e_3 ($\diamond$) can be

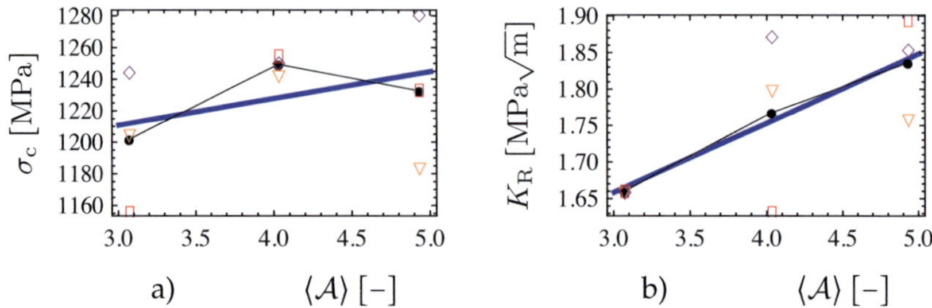

Figure 4.15: a) Maximum crack stress and b) maximum fracture resistance values for the three considered geometries under loading direction e_1 ($\triangledown$), e_2 ($\square$) and e_3 ($\diamond$), mean values ($\bullet$) and trend line due to linear regression (—)

addressed to the presence of a large and elongated grain, which is aligned mainly in this direction. The right side of Figure 4.16 a) shows the displacement field. The jump in the field is next to the grain, which indicates, that the crack has to propagate around this grain. The subfigure b) shows the corresponding stress field. In the region with the separation, the stress field is almost vanishing due to the post-failure unloading. The load is carried by the elastic bridging grain, where loads up to 2000 MPa are localized.

For this geometry (II) a considerable outlier is observed, when loaded in e_2-direction (Figure 4.15 ($\square$)). This decreased fracture stress and toughness has its origin in the presence of an unreinforced plane in the unit cell, which can be seen in Figure 4.16 c) and d). Here, the displacement and the stress field are depicted, after the fracture occurred. The fracture surface is extremely planar due to no bridging grains.

Summarizing, it can be said, that the results are affirmating the experimental observations, where long and elongated grains lead to higher fracture toughness. The result for the unit cell II has

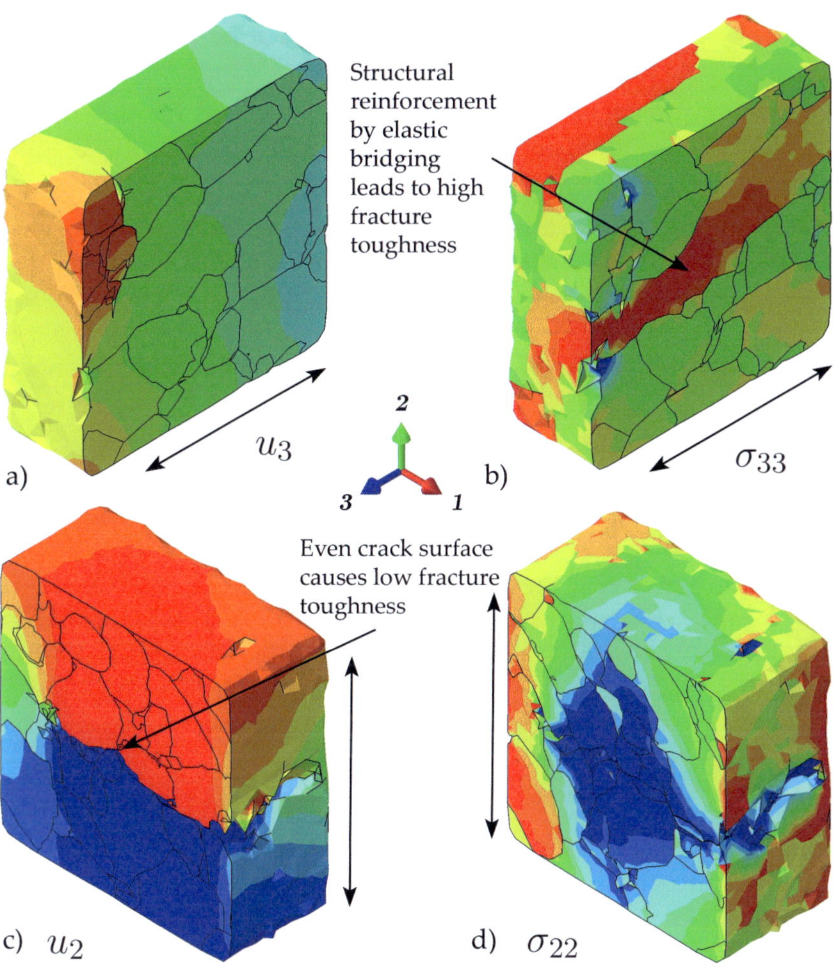

Figure 4.16: Displacement and stress fields for the unit cell *III*; a) Displacement and b) stress at the onset of fracture with bridging grain (loading direction e_3), c) Displacement and d) stress field after fracture without structural reinforcement (loading direction e_2)

shown what the effect of missing denticulation of the structure is: A significant decrease of the fracture toughness, but not of the strength.

Concerning the concrete values of the determined fracture toughnesses it has to be mentioned that they underestimate the fracture toughness, which can be observed experimentally for silicon nitride. This is not unexpected, due to the simplicity of the chosen fracture mechanics framework and due to the very limited size of the unit cube, which is beneath the critical crack length in silicon nitride. The procedure solely tried to give an idea over the fracture toughness on the microlevel. For this conceptual consideration, the results are confirming the picture, which is given by experimental observations of, e.g., Fett et al. (2008): The R-curve starts for very short cracks of single microns at a level, which is approximately one half of the plateau level, as it can be seen in Figure 4.14 d). The decrease afterwards can be addressed to the complete failure of the limited size of the unit cells. Therefore, the main features of the complex fracture process in silicon nitride can be captured by the unit cell approach with structure and the related constitutive assumptions.

4.5.3 Influence of Stress-Triaxiality

Stress fields in solids are in general inhomogeneous, which can be attributed to phase contrasts in the material on the microscale (Section 3.5.3) or to the geometry of the loaded body, e.g., notches or cracks. Those inhomogeneities can be quantified by triaxiality measures, which have been introduced earlier and will be considered in more detail later in this section. A well known fact is the influence of the stress triaxialities on the effective fracture behavior. A wide body of literature on the effects of triaxiality on the degradation of ductile materials is available. Prominent examples are the fundamental works of Gurson (1977) or Lemaitre (1985), where the pore growth or the formation of microcracks are identified as damaging mechanisms. Both concepts include the reduction of the internal load carrying capacity and the observation of ductile-to-brittle transition with increasing stress triaxiality $\mathcal{T} = -p/\sigma_{\mathrm{eq}}$. $\mathcal{T}$ is further tagged as 'classical' triaxiality. Here, the invariants of the stress tensor are considered only, which means that the influence of the ratio of the principal stresses itself do not enter the consideration. The consequence is that shear stresses are not considered as relevant for the degradation. Here, a different triaxiality measure, which has been introduced by Lode (1926), provides an option for doing so. It has to be emphasized that the triaxiality is mainly considered in the field of elastoplasticity. Nevertheless, an influence on a quasi-brittle material can be expected as well.

The following examinations are based on a parameterization of the stress tensor by the two triaxiality measures and the maximum principle stress, which allows for the determination of a corresponding deformation state. All calculations assume the effective isotropic elasticity of the bulk material at room temperature. This is the case only prior to fracture. The isotropic stiffness or compliance can be described by a consideration of the projector properties:

$$\bar{\mathbb{C}} = 3\bar{K}\mathbb{P}_1 + 2\bar{G}\mathbb{P}_2 \quad \Leftrightarrow \quad \bar{\mathbb{S}} = \frac{\mathbb{P}_1}{\bar{\mathbb{C}} \cdot \mathbb{P}_1} + \frac{\mathbb{P}_2}{\bar{\mathbb{C}} \cdot \mathbb{P}_2} = \frac{\mathbb{P}_1}{3\bar{K}} + \frac{\mathbb{P}_2}{2\bar{G}}. \tag{4.58}$$

Here, the values from Table 3.1 from the 238 grain unit cell for the bulk and shear modulus, $\bar{K} = 236.9$ GPa and $\bar{G} = 120.3$ GPa are used. So, an effective principle stress state, which is characterized by a certain effective triaxiality measure can be impressed on a unit cube by the usage of the parameterized effective stress tensor.

Classical Triaxiality. For the classical triaxiality a transverse-isotropic stress state can be assumed. Hence, only one degree of freedom is left, such that

$$\bar{\boldsymbol{\sigma}} = \bar{\sigma}_0 \left(\boldsymbol{B}_1 + k_{\mathcal{T}}(\boldsymbol{I} - \boldsymbol{B}_1) \right) \tag{4.59}$$

with $k_T < 1$ is used for the description of the stress state. Inserting this into the definition of the effective hydrostatic pressure $\bar{p}$ and the equivalent stress $\bar{\sigma}_{\mathrm{eq}}$ after von Mises delivers

$$\bar{p} = -\frac{1}{3}\mathrm{tr}(\bar{\sigma}) = \frac{1}{3}\bar{\sigma}_0(1+2k_T); \quad \bar{\sigma}_{\mathrm{eq}} = \sqrt{\frac{3}{2}}\|\bar{\sigma}+\bar{p}I\| = \bar{\sigma}_0\,(1-k_T) \quad (4.60)$$

and with it the parameterized classical triaxiality

$$T = \frac{1+2k_T}{3\,(1-k_T)}. \tag{4.61}$$

This can be rearranged, so that the triaxiality parameter

$$k_T = \frac{3\,T-1}{3\,T+2}. \tag{4.62}$$

Hence, the stress tensor can also be denoted by

$$\bar{\sigma} = \bar{\sigma}_0\left(B_1 + \frac{3\,T-1}{3\,T+2}(I - B_1)\right). \tag{4.63}$$

The limiting values $T > -2/3$ and $k_T < 1$ can be seen easily from this relations. Figure 4.17 a) shows the relationship between the triaxiality T and the parameter k_T.

The effective principle strains can be determined from the Hooke's law

$$\bar{\varepsilon} = \bar{\sigma}_0\frac{\bar{K}+\bar{G}T}{\bar{K}\bar{G}(2+3T)}\left(B_1 + \left(1 - \frac{3\bar{K}}{2\,(\bar{K}+\bar{G}T)}(I - B_1)\right)\right). \tag{4.64}$$

In Figure 4.17 b) the maximum $(-\,-)$ and minimum $(-\cdot-)$ principal strain amplitude for an expected fracture stress of $\bar{\sigma}_0 = 1000$ MPa are depicted. The singularity for $T \to -2/3$ can be observed, which would lead to an infinite stress state, see, Eq. 4.63, and is therefore not admissible. For $T \to \infty$ the spherical strain state $\bar{\varepsilon} = \bar{\sigma}_0/3\bar{K}\,I$

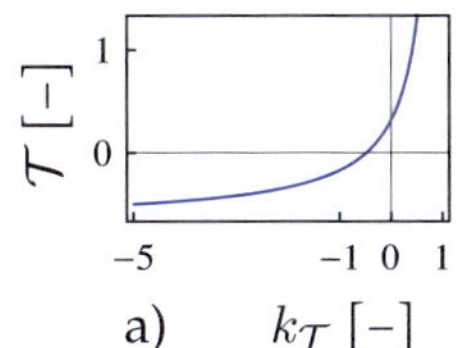 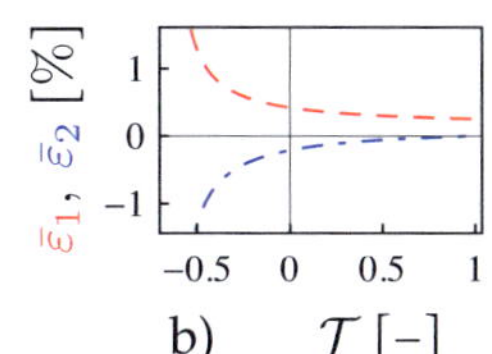 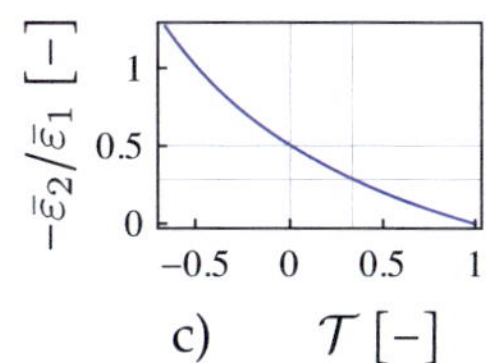

Figure 4.17: a) Triaxiality $\mathcal{T}$ as function of the parameter $k_\mathcal{T}$, b) maximum $(--)$ and minimum $(-\cdot-)$ principal strain amplitude and c) transverse contraction ratio as function of the triaxiality $\mathcal{T}$

is obtained. Interesting is the fact that the ratio of the principal strains, i.e., the negative of the scaling factor of the transverse principal strain in Eq. 4.64, is $1/2\,(1 - 1/\bar{\nu})$, when the engineering constants Young's modulus $\bar{E}$ and Poisson's ratio $\bar{\nu}$ are inserted. This remains finite for all $\bar{\nu} > 0$. For silicon nitride ($\bar{\nu} = 0.283$), it has a value of approximately 1.27 as it can be seen in Figure 4.17 c). Here, the value for $\bar{\nu}$ for the uniaxial tensile state with $\mathcal{T} = 1/3$ and the pure deviatoric deformation mode $\mathcal{T} = 0$ are marked.

The Lode Parameter. The parameterization of the effective principal stress and strain tensor with the Lode parameter

$$\mu = \frac{2\bar{\sigma}_2 - \bar{\sigma}_1 - \bar{\sigma}_3}{\bar{\sigma}_1 - \bar{\sigma}_3}, \quad \bar{\sigma}_1 \geqslant \bar{\sigma}_2 \geqslant \bar{\sigma}_3, \tag{4.65}$$

has two degrees of freedom, if the magnitude of the maximum principal stress is given by $\bar{\sigma}_1 = \bar{\sigma}_0$. Then $\bar{\sigma}_2 = \bar{\sigma}_0 k_2^\mu$ and $\bar{\sigma}_3 = \bar{\sigma}_0 k_3^\mu$. Hence, Lode's parameter ends in

$$\mu = \frac{2k_2^\mu - 1 - k_3^\mu}{1 - k_3^\mu}. \tag{4.66}$$

As it has been already mentioned in Section 3.5.3, general tensile stress states are characterized by $\mu = -1$, which is identical to $k_2^\mu = k_3^\mu$. Shear stresses are obtained for $\mu = 0$, which is characterized by $k_2^\mu = 1/2(1 + k_3^\mu)$. The compressive states $\mu = 1$ can be reached by $k_2^\mu = 1$. In Figure 4.18, the contours of Eq. (4.66) are depicted. For the parameters $k_2^\mu = k_3^\mu = 1$ the contour lines show indeterminacy, which is corresponding to a purely hydrostatic stress state.

For the further considerations it is interesting to see the influence of the Lode parameter μ on the fracture behavior, when a certain classical triaxiality $\mathcal{T}$ is given. So, the Eqs. (4.63) and (4.64) can be considered as a special case with $k_2^\mu = k_3^\mu$ or $\mu = -1$ of more general equations. The general parameterization can be developed, when the stress is denoted by

$$\bar{\sigma} = \bar{\sigma}_0 (\boldsymbol{B}_1 + k_2^\mu \boldsymbol{B}_2 + k_3^\mu \boldsymbol{B}_3). \tag{4.67}$$

Relatively compact expressions can be obtained, when the Lode parameter is introduced through the elimination of k_2^μ by Eq. (4.66), which gives the expression

$$\bar{\sigma} = \bar{\sigma}_0 \left(\boldsymbol{B}_1 + \frac{1}{2}(1 + k_3^\mu + \mu - k_3^\mu \mu)\,\boldsymbol{B}_2 + k_3^\mu \boldsymbol{B}_3 \right). \tag{4.68}$$

For this representation of the stress state, the classical triaxiality $\mathcal{T}$ can be determined, which allows for the elimination of the parameter k_3^μ and delivers

$$k_3^\mu = -\frac{3 + \mu - 3f_0}{3 - \mu + 3f_0}, \qquad f_0 = \mathcal{T}\sqrt{3 + \mu^2} \tag{4.69}$$

or inserted into the stress state

$$\bar{\sigma} = \bar{\sigma}_0 \left(\boldsymbol{B}_1 + \frac{2\mu - 3f_0}{3 - \mu + 3f_0}\,\boldsymbol{B}_2 - \frac{3 + \mu - 3f_0}{3 - \mu + 3f_0}\,\boldsymbol{B}_3 \right). \tag{4.70}$$

These equations yield a restriction on the admissible states for the classical triaxiality $\mathcal{T}$. Claiming a strictly positive denominator

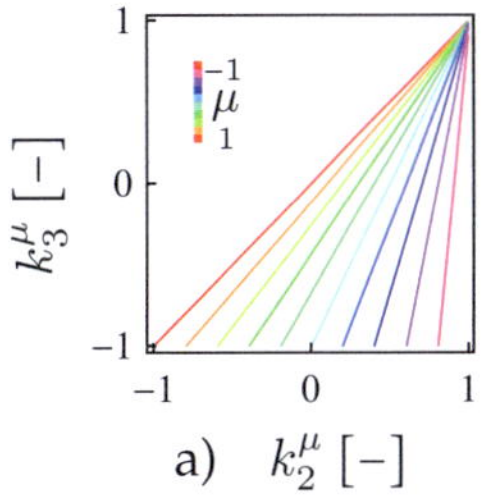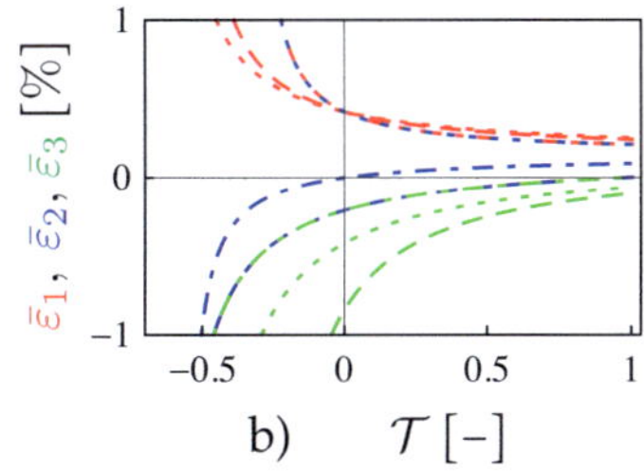

Figure 4.18: a) Lode's parameter contours as function of the parameterization with k_2^μ and k_3^μ and b) principal strains as function of the classical triaxiality for different the Lode parameters $\mu = -1$ $(-\cdot-)$, $\mu = 0$ $(---)$ and $\mu = 1$ $(\cdots)$

ends with

$$\mathcal{T} > \frac{\mu - 3}{3\sqrt{3 + \mu^2}}, \tag{4.71}$$

where the aforesaid limit $\mathcal{T} > -2/3$ for $\mu = -1$ is included. Like in Eq. (4.64), the effective principal strain tensor can be calculated to

$$\bar{\varepsilon} = \bar{\mathbb{S}}[\bar{\sigma}] = f_0^* (f_1\,\boldsymbol{B}_1 + f_2\,\boldsymbol{B}_2 + f_3\,\boldsymbol{B}_3), \tag{4.72}$$

where the coefficients

$$\begin{aligned}
f_0^* &= \bar{\sigma}_0 (2\bar{G}\bar{K}(3(1 + f_0) - \mu))^{-1}, &\tag{4.73}\\
f_1 &= 2f_0\bar{G} + (3 - \mu)\bar{K}, &\tag{4.74}\\
f_2 &= 2(f_0\bar{G} + \mu\bar{K}), &\tag{4.75}\\
f_3 &= 2f_0\bar{G} - (3 + \mu)\bar{K} &\tag{4.76}
\end{aligned}$$

have been introduced for sake of lucidity. The principal strain components are depicted in Figure 4.18 b). The trend is generally similar with Figure 4.17 b), however the different limiting triaxiality

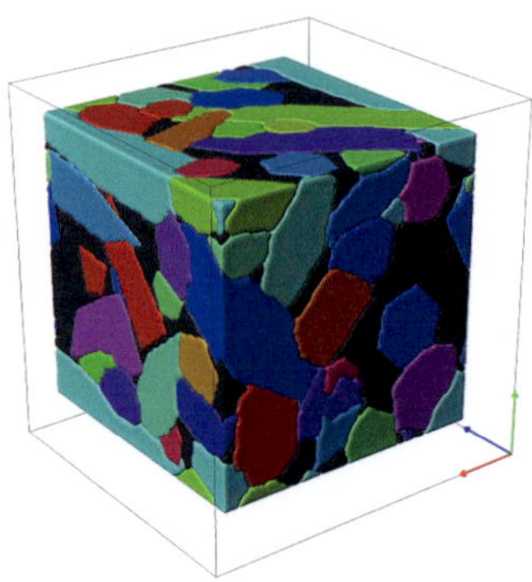

Figure 4.19: Unit cell with 36 grains used

values can be seen well pronounced. Interesting is the observation, that for the compressive state $\mu = 1$ ($\cdots$) the second principal strain is higher than the first principal strain, although the principal stress convention Eq. $(4.65)_2$ is a priori fulfilled.

For the determination of the triaxiality influence on the fracture stress of silicon nitride a unit cell, consisting of 36 grains with a mean aspect ratio of 3.7 and 88% grain volume fraction has been used (see, Figure 4.19).

The triaxiality $\mathcal{T}$ has been varied between 0 and $4/3$ in 5 intervals and Lode's parameter has been diversified between -1 and 1 in 7 levels, such that 35 simulations have been carried out and evaluated for the effective stress and strain curves. In Figure 4.20, two prominent examples of loading paths, i.e., volume average of stress over strain, are shown. The corner combinations lowest/highest Lode parameter and triaxiality. Both curves start at the strain level due to the thermal cool down. It can be seen, that the stress triaxiality affects all features of the brittle material behavior: The stiffness in the tensile direction, the fracture stress and degradation process. This can be attributed to the complex interactions of the stress fields in the grains and the tractions on the grain boundaries, which are

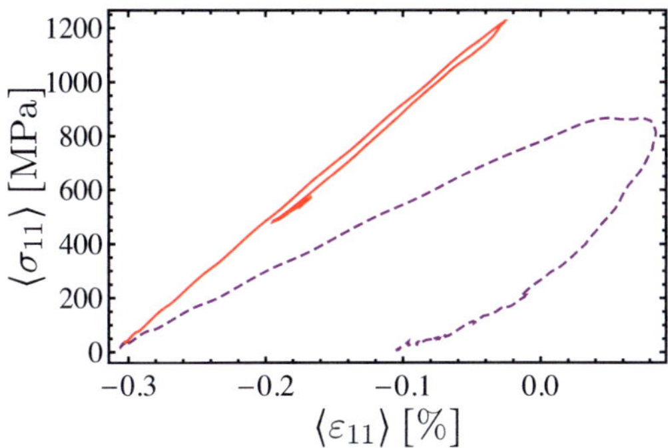

Figure 4.20: Volume averages of stress and strain for two combinations of Lode's parameter and stress triaxiality $\{\mu, \mathcal{T}\}$: $\{-1, 4/3\}$ (—) and $\{1, 0\}$ (– – –)

relevant for the microscopic and with it for macroscopic damage behavior. A loading with high triaxiality $\mathcal{T}$ under mainly tensile deformation delivers a significant steeper elastic slope and a higher fracture stress with an extremely brittle fracture afterwards (—). The unloading path is almost parallel to the loading path, which is an indication for few frictional contacts between the crack flanks. The case of high Lode parameter and low triaxiality (– – –) shows a different picture. Here, the elastic slope is relatively shallow and the fracture strength is significantly decreased. The unloading is characterized by a hysteresis due to the frictional dissipation, which occurs between the compressively loaded crack flanks.

In Figure 4.21 a) the results of all simulations are comprised. Here, the fracture stress is considered to be the stress-maximum of the effective stress-strain curve, because here final fracture occurs. It is clearly visible, that the resistance of the material is sensitive on both the Lode parameter and the triaxiality. The peak fracture stress level with value $\bar{\sigma}_c \approx 1250$ MPa is obtained at a high triaxiality and low Lode parameters, where a significant share of hydrostatic

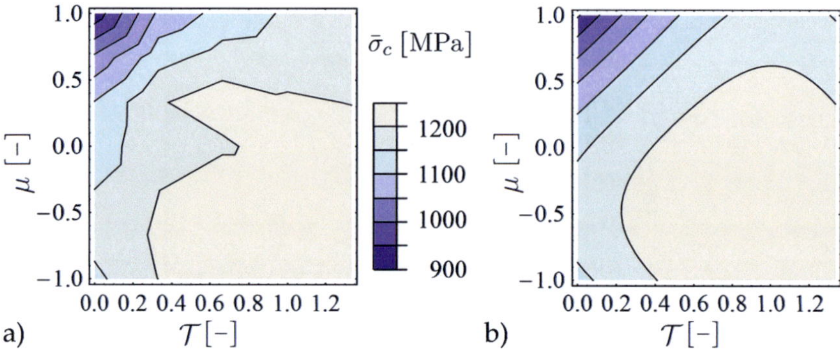

Figure 4.21: Effective fracture stress $\bar{\sigma}_\mathrm{c}$ in dependence of triaxiality $\mathcal{T}$ and Lode's parameter μ; a) results from the simulations b) bi-quadratic fit with the function $\bar{\sigma}_\mathrm{c}(\mathcal{T}, \mu) = \bar{\sigma}_\mathrm{c}^0 \, \boldsymbol{v}^{\mathcal{T}} \cdot \boldsymbol{A} \boldsymbol{v}^\mu$

stress due to hindered transverse contraction and negligible shear contribution avoid an early failure, which is abrupt then. The lowest values are obtained for a combination of low triaxiality and high Lode parameters. This can be interpreted, when Figure 4.18 b) is considered again. The corresponding deformation state can characterized by a biaxial tension in the e_1-e_2-plane and compression in e_3-direction. This leads to unfavorable combinations of tensile and shear stresses, which weakens the material especially on the interfaces but leads to a slower degradation process. On the other hand, a slower degradation process is observed, when the corresponding curve in Figure 4.20 (– – –) is viewed.

In subfigure b) the resulting fracture strengths are adjusted to the bi-quadratic function. This form is chosen due to the expected, slightly nonlinear influence of the triaxiality measures. Higher order estimates are possible, but are not expected to deliver further insight. Hence, the fracture strength be written as

$$\bar{\sigma}_{\mathrm{c}}(\mathcal{T},\mu) = \bar{\sigma}_{\mathrm{c}}^{0}\, \boldsymbol{v}^{\mathcal{T}} \cdot \boldsymbol{A} \boldsymbol{v}^{\mu}, \quad \boldsymbol{v}_{\mathcal{T}} = \begin{pmatrix} 1 \\ \mathcal{T} \\ \mathcal{T}^{2} \end{pmatrix}, \quad \boldsymbol{v}_{\mu} = \begin{pmatrix} 1 \\ \mu \\ \mu^{2} \end{pmatrix} \tag{4.77}$$

with the polynomial vectors $\boldsymbol{v}^{\mu}$ and $\boldsymbol{v}^{\mathcal{T}}$ and the coupling matrix $\boldsymbol{A}$. The offset cracking stress for $\mathcal{T} = \mu = 0$ is $\bar{\sigma}_{\mathrm{c}}^{0} = 1140$ MPa. The coupling factors are

$$\boldsymbol{A} = \begin{pmatrix} 100.0 & -10.5 & -11.0 \\ 17.5 & 14.3 & 14.5 \\ -9.4 & -6.6 & -6.3 \end{pmatrix} \%. \tag{4.78}$$

The fact that the relative error between adjusted function and the underlying data is never greater than 3.2% and mostly beneath 1% appears the chosen form feasible.

4.5.4 Effective Material Behavior

Analytical Example – Double Cantilever Beam

A double cantilever beam is a very simple, however, realistic example for a fracture experiment under fracture mode I. The big advantage of this geometry is that an analytical solution for the energy release rate and, hence, the stress intensity factor can be derived by a standard procedure from the beam theory. The geometry in Figure 4.22 a) can be assumed to consist of two cantilever beams with free bending length a, which act as series connection. The system is assumed to be loaded by dead loads with the magnitude F. Hence, the compliance of the structure is $S = 4a^{3}/Ebh^{3}$. The total potential as sum of the elastic stored energy and the potential of the external forces F is given by $\Pi^{\mathrm{tot}} = -1/2\,F^{2}S$. For displacement boundary conditions, the potential of the external load vanishes, which delivers an a priori negative energy release rate and, therefore, stable

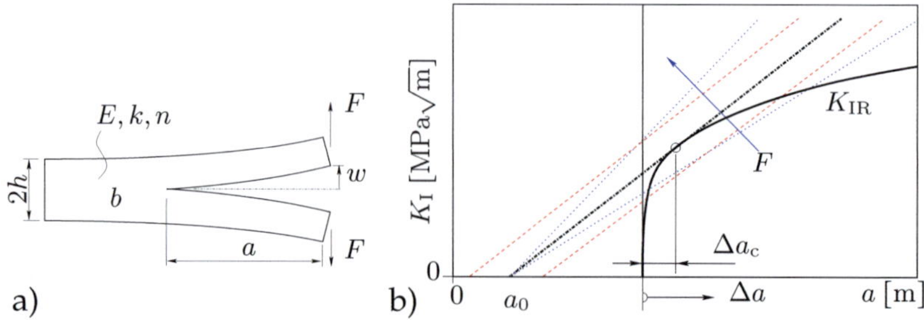

Figure 4.22: a) Double cantilever beam with geometric and constitutive quantities, b) *R*-curve behavior of the double cantilever beam for different initial crack lengths and applied loads

crack growth. For the dead load, the applied energy release rate can be derived by

$$\mathcal{G} = -\frac{1}{b}\frac{d\Pi^{\text{tot}}}{da} = \frac{12a^2 F^2}{Eb^2 h^3}.$$ (4.79)

Figure 4.22 b) shows the relationship for the stress intensity factors, which can be derived from the energy release rate by Eq. (4.56).

Crack propagation takes place, when the critical energy release rate is exceeded. The distinction can be made, when the tangent condition

$$\mathcal{G}(a) = \mathcal{G}_{\text{c}}(\Delta a) \quad \text{and} \quad \frac{\partial \mathcal{G}(a)}{\partial a} = \frac{\partial \mathcal{G}_{\text{c}}(\Delta a)}{\partial \Delta a}$$ (4.80)

for a constant force F is examined. The differential part of the equation system only comes into play, if the fracture resistance is not constant for increasing crack lengths $a = a_0 + \Delta a$. In order to visualize this effect, a simple power-law form for the critical energy release rate is assumed:

$$\mathcal{G}_{\text{c}} = k\Delta a^{1/n}.$$ (4.81)

With this assumption, the critical crack length increment is

$$\Delta a_{\mathrm{c}} = \frac{a_0}{2n - 1} \tag{4.82}$$

and the critical (and constant) force is

$$F_{\mathrm{c}} = \frac{1}{4\sqrt{3}} \frac{b}{n} \sqrt{E\,h^3\,k}\,\Delta a_{\mathrm{c}}^{\frac{1-2n}{2n}} . \tag{4.83}$$

This is an interesting result, because, it shows, that for a given geometry and the chosen R-curve behavior and all other properties known the critical crack length depends only on the initial crack length and the R-curve exponent. The like holds for the force. Different forces ($\cdots$) or initial crack lengths ($---$) lead to critical crack growth, when the tangent conditions in Eqs. (4.80) are fulfilled or the system has no solution. Otherwise, stable crack growth takes place.

Application of the Effective Damage Model on the Four-Point Bend Test

The implementation of the effective material behavior after Section 4.4 tries to describe the macroscopic material behavior by a very limited number of material parameters: the initial effective elastic isotropic constants, the effective initial crack stress $\bar{\sigma}_{\mathrm{c}}$ and the degradation velocity constant $\bar{H}$. Those parameters have to be chosen according to the information, which has been gathered from the microscopic simulations. This means, a unit cube, consisting of the effective material should show the same behavior as the underlying heterogeneous unit cube and have been determined to $800\ldots1200$ MPa for the effective fracture stress $\bar{\sigma}_{\mathrm{c}}$ and to $200\ldots390$ for the effective degradation velocity parameter $\bar{H}$. Note that for those high values between, the limiting relation Eq. (4.52) has to be evaluated.

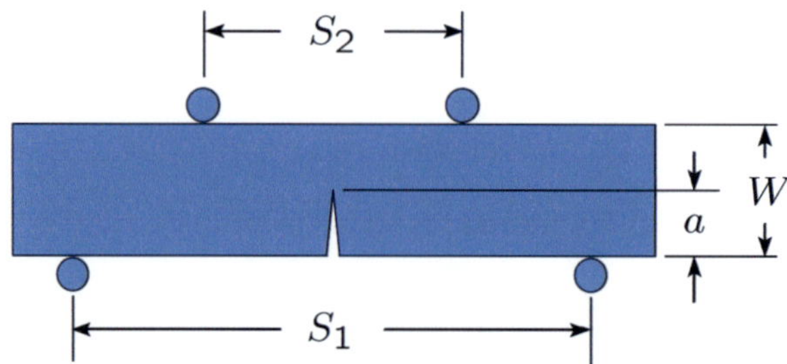

Figure 4.23: Four-point bending test with sharp notched specimen, see, e.g., Munz and Fett (1999)

As it has been already mentioned in Section 4.5.2, the crack resistance curve of a material and for a certain geometry can be determined from the load-displacement curve.

In order to achieve good comparability with existent experimental results, (see, Fett et al., 2008; Fünfschilling, 2010) the four-point bend test with sharp notched bars, according to DIN CEN/TS 14425-5 (2004), has been used to examine the effective fracture behavior (Figure 4.23). Important for the consideration of the steeply rising R-curve of structural reinforced ceramics is the onset of fracture, i.e., extremely small crack lengths. Therefore, a very small notch root radius of 7 µm according to Fett et al. (2008) has been chosen for the simulations, which requires a distinction between a long- and short-crack solution for the stress intensity factor.

Due to the sophisticated evaluation of the load-displacement curves, a short summary of the most important relations appears useful. First of all, the stress intensity factor for a long crack, i.e., a crack, which is significantly longer than the magnitude of the notch radius, is determined from the equation

$$K^* = \sqrt{\pi a}\,\sigma Y(\alpha) = \frac{F}{B\sqrt{W}}\hat{Y}(\alpha) \tag{4.84}$$

with the dimensionless crack length $\alpha = a/W$. The crack length can be computed from the compliance S of the specimen, because on the one hand it is known from the experiment $S = u/F$ and on the other hand it can be determined (semi)-analytically by a consideration of the geometry of the specimen. For a notched four-point bend bar the expression

$$\Delta S^* = \frac{9}{2}\pi \frac{(S_1 - S_2)^2}{BW^4 E'} \int_{a_0}^{a_0 + \ell} \hat{Y}(\tilde{a}/W)^2 \tilde{a}\, \mathrm{d}\tilde{a} \tag{4.85}$$

can be found. The semi-analytical geometry function $\hat{Y}$ is given in Fett and Munz (1997) by

$$\hat{Y} = \frac{k_1^S}{(1-\alpha)^{3/2}} \left(15 - 10\alpha + 3\alpha^2 + 120\alpha^2(1-\alpha)^6 + 9\exp\left(-\frac{k_2^S \alpha}{1-\alpha}\right) \right), \tag{4.86}$$

with the numerical factors $k_1^S = 4.67\cdot 10^{-2}$ and $k_2^S = 6.1342$.

The short-crack solution for the stress intensity factor is given by the relation from Munz and Fett (1997)

$$\frac{K}{K^*} \cong \tanh\left(k^* \sqrt{\frac{\ell}{R}} \right), \qquad k^* = 2.243. \tag{4.87}$$

Here, the special considerations of weight functions allows for a distinction between the influence of the notch and the crack. Following (Fett, 1995), this factor has to be included into the integrand of Eq. (4.85) (Fett, 1995), such that

$$\Delta S = \frac{9}{2}\pi \frac{(S_1 - S_2)^2}{BW^4 E'} \int_0^\ell \hat{Y}\left(\frac{a_0 + \tilde{\ell}}{W}\right)^2 \tanh^2\left(k^* \sqrt{\frac{\ell}{R}} \right) \tilde{\ell}\, \mathrm{d}\tilde{\ell} \tag{4.88}$$

is obtained as short crack solution for the compliance increment.

The simulations have been carried out on a model with extremely refined mesh in the notch root, in order to be able to resolve the intrinsic length scale of the crack propagation. The bending specimen has been deformed quasi-statically by a pure Dirichlet boundary conditions. The flexure of the bend bar is measured in the middle of the sample. Exemplarily, a force-displacement curve is depicted in Figure 4.24. The black dots indicate the finite element solution, which shows a slight scattering due to the material instabilities. This leads to significant influence on the determination of the crack length and the fracture resistance. In order to obtain a smooth solution, a linear fit used. Following Fünfschilling (2010), the compliance increment from the test is given by

$$\Delta \hat{S} = \frac{u - S_0 \, F}{F}.$$
(4.89)

It is used for the determination of the crack length ℓ by solving the equation $\Delta \hat{S}(F, u) = \Delta S(\ell)$ with Eq. (4.88) numerically. Once, the crack length is determined, the R-curve can be calculated by the Eqs. (4.84) and (4.87).

For the visualization of the influence of the effective degradation rate $\bar{H}$ on the R-curve behavior of the model, five simulations with an assumed effective fracture strength $\bar{\sigma}_c = 1000$ MPa and the values $\bar{H} = \{100, 150, 200, 250, 390\}$ have been carried out. The results are compiled in Figure 4.25. The subfigure a) shows the adjusted force-displacement curves from the simulations together with the initial slope, i.e., the elastic response of the sample without fracture behavior. The difference between the curves appears comparable small, but a significant effect of the stiffness degradation constant $\bar{H}$ is noticed, when the applied force is plotted over the compliance increment $\Delta \hat{S}$, see, subfigure b). The crack lengths, based on the small-crack solution are depicted in subfigure c). As expected, the cracks in the specimen are growing faster, when the degradation constant $\bar{H}$ increases. In subfigure d) the short-crack solution of the R-curve is plotted. The resulting curve for the highest value of $H = 390$ is reproducing the results of Fett et al. (2008). However, a

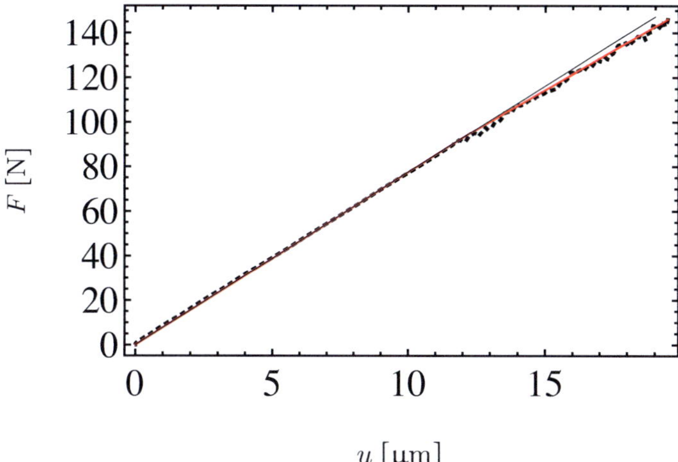

Figure 4.24: Force over displacement, finite element solution ($\cdots$), initial slope (—) and linear fit of the fracture load path (—)

slight overestimation of the plateau value can be observed as well. It has to be noticed that the model is used at its conceptual limit, due to the restrictions for the choice of the degradation parameter, see, Eq. (4.52). However, the suggested procedure is capable of capturing the microscopic fracture behavior and allows for a usage of the obtained results for the effective behavior.

4.6 Conclusions

The fracture behavior of silicon nitride has been described by micromechanical finite element simulations, which have been used to address the different aspects of the complex behavior on the microscale. Hence, the results on the microstructure (Chapter 2) and the thermoelasticity (Chapter 3) have been incorporated into a more general context, which includes fracture due to phase cracking and

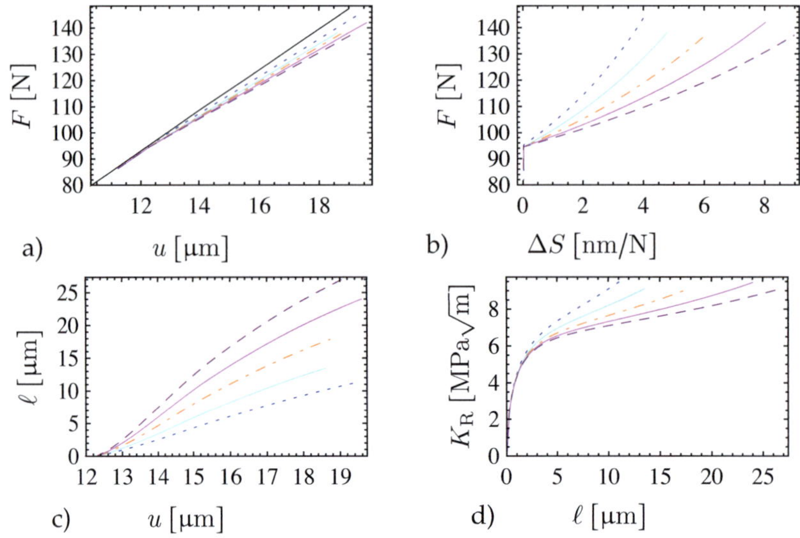

Figure 4.25: Fracture evolution of the four-point bend test; initial slope (—), $\bar{H} = 100$ ($\cdots$), $\bar{H} = 150$ (—), $\bar{H} = 200$ ($- \cdot -$), $\bar{H} = 250$ (—), $\bar{H} = 390$ ($---$); a) force over displacement, b) force over compliance increment, c) crack length over displacement, d) fracture resistance over crack length

interface delamination. Based on the unit cell simulations, an averaging technique allows for an estimation of the effective material behavior.

For a thermodynamical consistent unit cell approach, it is a precondition that the consistency of microscopic and macroscopic stress power is preserved. In case of purely thermoelastic behavior, this is a priori fulfilled, because the Hill-Mandel condition holds for the chosen boundary conditions. In the case of fracture due to interface delamination the Hill-Mandel condition had to be extended in order to capture the contributions of the tractions and separations. The concept has been proved by a consideration of a unit cube, which has been exposed to a shear deformation. Here, the consistency between micro- and macroscopic stress power for the different features of the implemented material behavior, i.e., elasticity, fracture and post-failure friction could be observed, such that general consistency can be attributed.

Furthermore, the chapter was focused on a detailed consideration of the fracture behavior of the solid phases and the phase boundaries. Models for fracture under maximum principal stresses for the isotropic glassy matrix and the assumption of anisotropic fracture of the β-grains have been suggested. The used fracture strengths have been motivated by experimental observations and ab-initio simulations. The grain boundary delamination has been implemented as an extension of Coulomb's friction law with respect to elasticity and degradation. For this, a thermodynamical consistent framework has been used. In particular, the dissipation due to interface debonding has been maximized under two side conditions, i.e., the delamination criteria for tensile and for shear delamination. The equations have been examined in detail, such that an upper limit for the choice of the degradation parameters could be found by a consideration of the interface tangent operator.

All mentioned features have been incorporated into a simulation framework for the usage with Abaqus/Explicit.

The unit cell approach has been used to improve the understanding of the influence of the microstructure on the effective properties. In order to do so, three different cells with mean aspect ratios between 3 and 5 have been created and loaded in different spatial directions. The obtained effective stress-strain curves have been evaluated by the fundamental equations of the linear elastic fracture mechanics, which delivered the fracture toughness of the material. A clear trend towards increasing fracture toughness with increasing mean aspect ratio can be attributed, although a certain influence of the microstructure has been observed, as well. This is consistent with the experimental observations on silicon nitride samples and can be addressed to the load carrying capacity of large and elongated grains. So, it was observed that the fracture toughness is decreased significantly, when the crack can propagate in a plane, where no interlocking grains are providing elastic bridges as crack obstacle. On the other hand, the enormous effect of the elastic bridging grains could be observed in detail.

Furthermore, the influence of stress triaxiality has been examined, in order to understand the effect of inhomogeneous loads on the brittle material. In order to do so, the principal stress tensor has been parameterized by the maximum principal stress, the classical stress triaxiality and Lode's parameter. The obtained stress tensor has been used to derive an effective strain tensor, which delivered the prescribed displacement boundary conditions. Findings are that the fracture stress is significantly increased, when the material is loaded by high stress triaxiality and mainly tensile stress fields, i.e., a low Lode parameter. The opposite relation holds as well. The detailed evaluation of fracture stresses allowed for the empirical adjustment of the relationship between fracture stresses, classical triaxiality and the Lode parameter.

Hence, it can be concluded, that it is in generally possible to capture the main features of silicon nitride in finite element simulations. In particular, the most important feature for the high fracture toughness and the steeply rising R-curve has been carefully examined. It can be addressed to the existence of grains, which are forming elastic bridges over the cracks faces on the microscopic scale. Here, the highest fracture toughnesses could be observed, when the grains where loaded until the assumed fracture stress has been exeeded and the grain cracked. This was mainly the case for high mean aspect ratios and high stress triaxialities. The fracture strength and toughness has been decreased significantly, when the fracture occurred only intergranular, which was the case for low aspect ratios and biaxial tensile loading states with uniaxial compression or shear dominance.

On the other end, the effective fracture properties of silicon nitride have been described by a damaging material, which assumes anisotropic degradation of strength and stiffness in the direction of the maximum principal stress. Like the interface behavior, the effective fracture behavior has been implemented by the notion of dissipation due to material degradation. An upper limit for the choice of the degradation constants has been derived analytically from the continuum tangent operator. The parameters of the effective material have been determined from the results of the unit cell simulations. The obtained effective material behavior has been used to describe the fracture of four-point bend bars. From the achieved load-displacement curves the determination of the corresponding *R*-curves was possible. Here, the short-crack solution of Fett et al. (2008) delivered results, which are in close correspondence with the experimental observations.

Concluding, the statement should be allowed that the whole framework is delivering excellent results, which prove the concept of structural reinforcement by elastic bridging grains. These findings explicitly support the results of Fünfschilling et al. (2011) and, therefore, are describing one of the most important material features of silicon nitride for structural applications.

Chapter 5

Discussion and Outlook

The following discussion chapter tries to step back and take a look at the whole picture. In order to avoid boredom of the reader, a detailed summery and conclusions on the particular considerations of this work are given at the end of each specific chapter.

Summing up the entire work, it can be said that the limits for finite element simulations on the microscale have been pushed forward considerably. The concentration of elaborated techniques into one simulation framework is unique in the field to the best knowledge of the author.

Hence, it might be interesting to start with a short consideration of the concept in general. The focus of this work lies on the simulation of a highly complex material behavior on the microstructural level and the subsequent determination of effective properties. Due to the key relevance of the microgeometry as well as the microbehavior a huge amount of details had to be considered to be able to implement the concept, which has been introduced in Figure 1.2. The highest priority was the experimental motivation and mathematically strict implementation of the concept. So, experimental observation have been used as direct as possible for the implementation of the microstructure, thermoelastic, and fracture behavior. Of course, it has to be admitted that this often was difficult due to a lack of relevant data. At some points a good portion of intuition combined with all the necessary rigor has lead to justifiable assumptions and approximations. Examples are the case consideration for the implementation of the structure generator, the introduction of the projected periodic boundary conditions, implementation of the interface behavior, fracture behavior of the phases, and, especially, the homogenization of the complex behavior.

In the end, the concept could be proved by a comparison of experimental results with numerical simulations on the macroscale, which were based on the results of the microscopic simulations.

The most important conclusion for the research on silicon nitride is that the grain bridging is of crucial importance. Of course, this is not essentially new, but due to the fact, that the fracture behavior on the microscale is experimentally not observable, the simulations deliver significant proof to the concept of the elastic bridging grains and with it to the concept of the related bridging stresses.

The key achievement of this work is the practical implementation of a numerical framework, which is straight forward, thermodynamically consistent and – within its field of application – efficient. It appears feasible, to apply the used techniques to any other brittle material through the usage of different geometries, e.g., a Voroni or a Johnson-Mehl tessellation would meet the requirements for a geometrical representation of alumina or zirconia. The same holds for the material assumptions for thermoelasticity and the fracture assumptions. An extension of the material and interface assumption into the domain of plasticity would open up the application range for cermets or hard metals like tungsten.

The perspective of the created microscale simulation framework is an application for further research on silicon nitride. Interesting questions that might be addressed, include the influence of different grain boundary toughnesses for a given geometry or the influence of the temperature on the fracture process. A consideration of wear could be enlightening as well.

A wide range of potential applications of the model for the elastic degradation of the effective material is possible as well. So, most of the ceramics and other non-metals like concrete or rock fail under fracture modes, which allow for the assumption of damage.

Technical improvement of the model can be achieved at several points as well. It would be highly desirable to have the option of perfectly periodic finite element meshes, which would make the usage of a Robin boundary condition superfluous in the fracture simulations. The fracture of the phases could be significantly improved by applying the extended finite element method. Here, numerical artifacts due to the deletion of elements could be avoided.

Finally, one can say that the implementation of the finite element framework for fracture simulations of microheterogeneous material has been a great success and should be widely used for the improvement of the understanding of complex microstructure-property relations, which can influence the development of new or at least better materials.

Bibliography

Ahrens, J., Kohrt, K., 1983. ALGORITHM 599 – Sampling from Gamma and Poisson distributions. ACM Transactions on Mathematical Software 9/2, 255 – 57.

Bažant, Z., Belytschko, T., 1985. Wave propagation in a strain softening bar; exact solution. Journal of Engineering Mechanics 111, 381 – 89.

Becher, P., Sun, E., Plucknett, K., Alexander, K., Hsueh, C., Lin, H.-T., Waters, S., Westmoreland, C.-G., 1998. Microstructural design of silicon nitride with improved fracture toughness I: Effects of grain shape and size. Journal of Crystal Growth 81, 2821 – 30.

Böhlke, T., Bertram, A., 2001. The evolution of Hooke's law due to texture development in FCC polycrystals. International Journal of Solids and Structures 38, 9437 – 9459.

Böhlke, T., Brüggemann, C., 2001. Graphical representation of the generalized Hooke's law. Technische Mechanik 21/2, 145 – 58. URL `http://preview.tinyurl.com/6e9uxkc`

Borbély, A., Kenesei, P., Biermann, H., 2006. Estimation of the effective properties of particle-reinforced metal-matrix composites from microtomographic reconstructions. Acta Materialia 54, 2735 – 44.

Cooper, D., 1988. Random-sequential-packing simulations in three dimensions for spheres. Physical Review A 38, 522 – 24.

Costanzo, F., Boyd, J., Allen, D., 1996. Micromechanics and homogenization of inelastic composite materials with growing cracks. Journal of Mechnics and Physics of Solids 44, 333 – 40.

DIN CEN/TS 14425-5, 2004. Hochleistungskeramik-Prüfverfahren zur Bestimmung der Bruchzähigkeit von monolithischer Keramik – Teil 5: Verfahren für Biegeproben mit V-Kerb (SEVNB-Verfahren).

Federov, F., 1968. Theory of Elastic Waves in Crystals. Plenum Press, New York.

Fett, T., 1995. Notch effects in determination of fracture toughness and compliance. International Journal of Fracture 72, 27 – 30.

Fett, T., Fünfschilling, S., Hoffmann, M., Oberacker, R., 2008. *R*-curve determination for the initial stage of crack extension in Si_3N_4. Journal of the American Ceramic Society 91, 3638 – 42.

Fett, T., Munz, D., 1997. Stress Intensity Factors and Weight Functions. Computational Mechanics Publications, Southampton, UK.

Fett, T., Munz, D., Kounga Njiwab, A., Rödel, J., Quinn, G., 2005. Bridging stresses in sintered reaction-bonded Si_3N_4 from COD measurements. Journal of the European Ceramic Society 25, 29 – 36.

Flaquer, J., Rios, A., Martín-Meizoso, A., Nogales, S., Böhm, H., 2007. Effect of diamond shapes and associated thermal boundary resistance on thermal conductivity of diamond-based composites. Computational Materials Science 41 (2), 156 – 163.

Fritzen, F., Böhlke, T., 2010. Periodic three-dimensional mesh generation for particle reinforced composites with application to metal matrix composites. International Journal of Solids and Structures 48, 706 – 18.

Fritzen, F., Böhlke, T., Schnack, E., 2009. Periodic three-dimensional mesh generation for crystalline aggregates on Voronoi tesselations. Computational Mechanics 43, 701 – 713.

Fünfschilling, S., 2010. Mikrostrukturelle Einflüsse auf das R-Kurvenverhalten bei Siliciumnitridkeramiken. Ph.D. thesis, KIT - Campus South.

Fünfschilling, S., Fett, T., Hoffmann, M., Oberacker, R., Schwind, T., Wippler, J., Böhlke, T., Özcoband, H., Schneider, G., Becher, P., Kruzic, J., 2011. Mechanisms of toughening in silicon nitrides: The roles of crack bridging and microstructure. Acta Materialia 59, 3978 – 89.

Govindjee, S., Kay, G., Simo, J., 1995. Anisotropic modelling and numerical simulation of brittle damage in concrete. International Journal for Numerical Methods in Engineering 38, 3611 – 33.

Gross, D., Seelig, T., 2007. Bruchmechanik. Springer Berlin Heidelberg.

Gurson, A., 1977. Continuum theory of ductile rupture by void nucleation and growth, part I: Yield criterion and flow rules for porous ductile materials. Ph.D. thesis, Brown University.

Gustafson, K., Abe, T., 1998. The third boundary condition – was it Robin's? The Mathematical Intelligencer 20, 63 – 71.

Hampshire, S., Nestor, E., Flynn, R., Goursat, P., Sebai, M., Thompson, D., Liddell, K., 1994. Yttrium oxynitride glasses: Properties and potential for crystallisation to glass-ceramics. Journal of the European Ceramic Society 14, 261 – 73.

Henderson, C., Taylor, D., 1975. Thermal expansion of the nitrides and oxynitrides of silicon in relation to their structures. Transactions and Journal of the British Ceramic Society 74, 49 – 53.

Hibbitt, Karlsson, Sorensen, 2001. ABAQUS/CAE user's manual. Hibbitt, Karlsson & Sorensen, Inc.

Hill, R., 1963. Elastic properties of reinforced solids: Some theoretical principles. Journal of the Mechanics and Physics of Solids 11, 357 – 72.

Iba, H., Naganuma, T., Matsumara, K., Kagawa, Y., 1999. Fabrication of transparent continuous oxynitrid glass fiber-reinforced glass matrix composit. Journal of Material Science 34, 5701 – 05.

Kinsler, L., Frey, A.R. Coppens, A., J.V., S., 2000. Fundamentals of Acoustics. John Wiley & Sons, Inc., New York.

Kitayama, M., Hiraro, K., Toriyama, M., Kanzaki, S., 1998a. Modeling and simulation of grain growth in Si_3N_4 – I. Anisotropic ostwald ripening. Acta Metall. 46/18, 6541 – 50.

Kitayama, M., Hiraro, K., Toriyama, M., Kanzaki, S., 1998b. Modeling and simulation of grain growth in Si_3N_4 – II. The α-β transformation. Acta Metall. 46/18, 6551 – 57.

Kitayama, M., Hiraro, K., Toriyama, M., Kanzaki, S., 2000. Modeling and simulation of grain growth in Si_3N_4 – III. Tip shape evolution. Acta Materialia 48, 4635 – 40.

Krämer, M., Hoffmann, M., Petzow, G., 1993. Grain growth kinetics of Si_3N_4 during α/β-transformation. Acta Metallurgica 41/10, 2939 – 47.

Krämer, M., Wittnüss, D., H., K., Hoffmann, M., Petzow, G., 1994. Relations between crystal structure and growth morphology of β-Si_3N_4. Journal of Crystal Growth 140, 157 – 66.

Kruzic, J., Satet, R., Hoffmann, M., Cannon, R., Ritchie, R., 2008. The utility of *r*-curves for understanding fracture toughness-strength relations in bridging ceramics. Journal of the American Ceramic Society 91, 1986 – 94.

Lange, F., 1973. Relation between strength, fracture energy, and microstructure of hot-pressed Si_3N_4. Journal of The American Ceramic Society 56, 518 – 22.

Le Bourhis, E., 2008. Glass – Mechanics and Technology. Wiley-VHC.

Lemaitre, J., 1985. A continuous damage mechanics model for ductile fracture. Journal of Engineering Materials Technology 107, 83 – 89.

Lode, W., 1926. Versuche über den Einfluss der mittleren Hauptspannung auf das Fließen der Metalle Eisen, Kupfer und Nickel. Zeitschrift der Physik 36, 913 – 19.

Louis, P., Gokhale, A., 1996. Computer simulations of spatial arragement and connectivity of particles in three-dimensional microstructure: Application to model electrical conductivity of polymer matrix composite. Acta Materialia 44, 1519 – 28.

Lube, T., Dusza, J., 2007. A silicon nitride reference material – a testing program of esis tc6. Journal of the European Ceramic Society 27, 1203 – 09.

Maugin, G., 1992. The Thermomechanics of Plasticity and Fracture. Cambrigde University Press.

Mücklich, F., Hartmann, S., Hoffmann, M., Schneider, G., Ohser, J., Petzow, G., 1994. Quantitative description of Si_3N_4 microstructures. Key Engineering Materials 5, 465 – 70.

Munz, D., 07. What can we learn from r-curve measurements. Journal of the American Ceramic Society 90, 1 – 15.

Munz, D., Fett, T., 1997. Stress Intensity Factors and Weight Functions. Computational Mechanics Publications, Southampton, UK.

Munz, D., Fett, T., 1999. Ceramics – Mechanical Properties, Failure Behaviour, Materials Selection. Springer.

Nipp, K., Stoffer, D., 2002. Lineare Algebra. VDF Hochschulverlag AG.

Ogata, S., Hirosaki, N., Shibutani, Y., 2004. A comparative ab initio study of the 'ideal' strength of single crystal α- and β- Si_3N_4. Acta Materialia 52, 233 – 38.

Ohji, T., Hirao, K., Kanzaki, S., 1995. Fracture resistance behavior of highly anisotropic silicon nitride. Journal of the Amermican Ceramic Society 78, 3125 – 28.

Ohser, J., Mücklich, F., 2000. Statistical Analysis of Microstructures in Materials Science. Wiley-VHC.

Peillon, F., Thevenot, F., 2002. Microstructural designing of silicon nitride related to toughness. Journal of the European Ceramic Society 22, 271 – 78.

Peterson, I., Tien, T.-Y., 1995. Effect of the grain boundary thermal expansion coefficient on the fracture toughness in silicon nitride. Journal of the American Ceramic Society 78, 2345 – 52.

Satet, R., 2002. Einfluss der Grenzflächeneigenschaften auf die Gefügeausbildung und das mechanische Verhalten von Siliciumnitrid-Keramiken. Ph.D. thesis, Universität Karlsruhe (TH).

Shoemake, K., 1992. Graphics Gems III. Academic Press, London, Ch. Uniform Random Rotations, pp. 124 – 32.

Silverman, B., Jones, M., Wilson, J., Nychka, D., 1990. A smoothed EM approach to a class of problems in image analysis and integral equations. Journal of the Royal Statistical Society Series B 52, 271 – 324.

Simo, J., Hughes, T., 2000. Computational Inelasticity. New York, Springer.

Simo, J., Oliver, J., Armero, F., 1993. An analysis of strong discontinuities induced by strain-softening in rate-independent inelastic solids. Computational Mechanics 12, 277 – 96.

Sluys, L., 1992. Wave propagation, localization and dispersion in softening solids. Ph.D. thesis, Delft University of Technology.

Sun, E., Becher, P., Plucknett, K., C.-H., H., Alexander, K., S.B., W., 1998. Microstructural design of silicon nitride with improved fracture toughness II: Effects of yttria and alumina additives. Journal of the American Ceramic Society 81, 2831 – 40.

Suquet, P., 1982. Plasticité et homogénéisation. Ph.D. thesis, Université Pierre et Marie Curie, Paris.

Suquet, P., 1985. Local and global aspects in the mathematical theory of plasticity. pp. 279 – 310.
URL `http://preview.tinyurl.com/5whlysv`

Tschopp, M., Wilks, G., Spowart, J., 2008. Multi-scale characterization of orthotropic microstructures. Modelling and Simulation in Materials Science and Engineering 16, 1 – 14.

Vardi, Y., Shepp, L., Kaufman, L., 1985. A statistical model for positron emission tomography. Journal of the American Statistical Society 80 (389), 8 – 37.

Verhoosel, C., Remmers, J., Miguel, A., de Borst, R., 2010. Computational homogenization for adhersive and cohesive failure in quasi-brittle solids. International Journal for Numerical Methods in Engineering 83, 1155 – 79.

Vogelgesang, R., Grimsditch, M., Wallace, J., 2000. The elastic constants of single crystal β-Si$_3$N$_4$. Applied Physics Letters 76/8, 982 – 84.

Wei, Y., Anand, L., 2004. Grain-boundary sliding and separation in polycrystalline metals: application to nanocrystalline fcc metals. Journal of the Mechanics and Physics of Solids 52, 2587 – 616.

Wippler, J., Böhlke, T., 2010. Thermal residual stresses and triaxiality measures. In: PAMM – Proceedings in Applied Mathematics and Mechanics. Vol. 10. pp. 137 – 38.

Wippler, J., Böhlke, T., 2011. An algorithm for the generation of silicon nitride structures. Journal of the European Ceramics Society 32, 589 – 602.

Wippler, J., Fünfschilling, S., Fritzen, F., Böhlke, T., Hoffmann, M., 2011. Homogenization of the thermoelastic properties of silicon nitride. Acta Materialia 59, 6029 – 38.

Young, P., Beresford-West, T., Coward, S., Notarberardino, B., Walker, B., Abdul-Aziz, A., 2008. An efficient approach to converting three-dimensional image data into highly accurate computational models. Philosophical Transactions of the Royal Society A 366, 3155 – 73.

Schriftenreihe Kontinuumsmechanik im Maschinenbau
Karlsruher Institut für Technologie
(ISSN 2192-693X)

Herausgeber: Prof. Dr.-Ing. Thomas Böhlke

**Die Bände sind unter www.ksp.kit.edu als PDF frei verfügbar oder
als Druckausgabe bestellbar.**

Band 1 Felix Fritzen
**Microstructural modeling and computational homogenization
of the physically linear and nonlinear constitutive behavior of
micro-heterogeneous materials. 2011**
ISBN 978-3-86644-699-1

Band 2 Rumena Tsotsova
Texturbasierte Modellierung anisotroper Fließpotentiale. 2012
ISBN 978-3-86644-764-6

Band 3 Johannes Wippler
**Micromechanical Finite Element Simulations of Crack Propagation
in Silicon Nitride. 2012**
ISBN 978-3-86644-818-6